Selected Titles in This Series

713 **Dorina Mitrea, Marius Mitrea, and Michael Taylor,** Layer potentials, the Hodge Laplacian, and global boundary problems in nonsmooth Riemannian manifolds, 2001

712 **Raúl E. Curto and Woo Young Lee,** Joint hyponormality of Toeplitz pairs, 2001

711 **V. G. Kac, C. Martinez, and E. Zelmanov,** Graded simple Jordan superalgebras of growth one, 2001

710 **Brian Marcus and Selim Tuncel,** Resolving Markov chains onto Bernoulli shifts via positive polynomials, 2001

709 **B. V. Rajarama Bhat,** Cocylces of CCR flows, 2001

708 **William M. Kantor and Ákos Seress,** Black box classical groups, 2001

707 **Henning Krause,** The spectrum of a module category, 2001

706 **Jonathan Brundan, Richard Dipper, and Alexander Kleshchev,** Quantum Linear groups and representations of $GL_n(\mathbb{F}_q)$, 2001

705 **I. Moerdijk and J. J. C. Vermeulen,** Proper maps of toposes, 2000

704 **Jeff Hooper, Victor Snaith, and Min van Tran,** The second Chinburg conjecture for quaternion fields, 2000

703 **Erik Guentner, Nigel Higson, and Jody Trout,** Equivariant E-theory for C^*-algebras, 2000

702 **Ilijas Farah,** Analytic guotients: Theory of liftings for quotients over analytic ideals on the integers, 2000

701 **Paul Selick and Jie Wu,** On natural coalgebra decompositions of tensor algebras and loop suspensions, 2000

700 **Vicente Cortés,** A new construction of homogeneous quaternionic manifolds and related geometric structures, 2000

699 **Alexander Fel′shtyn,** Dynamical zeta functions, Nielsen theory and Reidemeister torsion, 2000

698 **Andrew R. Kustin,** Complexes associated to two vectors and a rectangular matrix, 2000

697 **Deguang Han and David R. Larson,** Frames, bases and group representations, 2000

696 **Donald J. Estep, Mats G. Larson, and Roy D. Williams,** Estimating the error of numerical solutions of systems of reaction-diffusion equations, 2000

695 **Vitaly Bergelson and Randall McCutcheon,** An ergodic IP polynomial Szemerédi theorem, 2000

694 **Alberto Bressan, Graziano Crasta, and Benedetto Piccoli,** Well-posedness of the Cauchy problem for $n \times n$ systems of conservation laws, 2000

693 **Doug Pickrell,** Invariant measures for unitary groups associated to Kac-Moody Lie algebras, 2000

692 **Mara D. Neusel,** Inverse invariant theory and Steenrod operations, 2000

691 **Bruce Hughes and Stratos Prassidis,** Control and relaxation over the circle, 2000

690 **Robert Rumely, Chi Fong Lau, and Robert Varley,** Existence of the sectional capacity, 2000

689 **M. A. Dickmann and F. Miraglia,** Special groups: Boolean-theoretic methods in the theory of quadratic forms, 2000

688 **Piotr Hajłasz and Pekka Koskela,** Sobolev met Poincaré, 2000

687 **Guy David and Stephen Semmes,** Uniform rectifiability and quasiminimizing sets of arbitrary codimension, 2000

686 **L. Gaunce Lewis, Jr.,** Splitting theorems for certain equivariant spectra, 2000

685 **Jean-Luc Joly, Guy Metivier, and Jeffrey Rauch,** Caustics for dissipative semilinear oscillations, 2000

684 **Harvey I. Blau, Bangteng Xu, Z. Arad, E. Fisman, V. Miloslavsky, and M. Muzychuk,** Homogeneous integral table algebras of degree three: A trilogy, 2000

(Continued in the back of this publication)

Layer Potentials, the Hodge Laplacian, and Global Boundary Problems in Nonsmooth Riemannian Manifolds

MEMOIRS
of the
American Mathematical Society

Number 713

Layer Potentials, the Hodge Laplacian, and Global Boundary Problems in Nonsmooth Riemannian Manifolds

Dorina Mitrea
Marius Mitrea
Michael Taylor

March 2001 • Volume 150 • Number 713 (fourth of 5 numbers) • ISSN 0065-9266

American Mathematical Society
Providence, Rhode Island

2000 *Mathematics Subject Classification.*
Primary 35J55, 42B20, 58J05, 58J32, 58A14; Secondary 31B10, 31C12, 45E05, 78A30.

Library of Congress Cataloging-in-Publication Data

Mitrea, Dorina (Dorina Irena Rita), 1965–
Layer potentials, the Hodge Laplacian, and global boundary problems in nonsmooth Riemannian manifolds / Dorina Mitrea, Marius Mitrea, Michael Taylor.
p. cm. — (Memoirs of the American Mathematical Society, ISSN 0065-9266 ; no. 713)
"Volume 150, number 713 (fourth of 5 numbers)
Includes bibliographical references.
ISBN 0-8218-2659-X
1. Riemannian manifolds. 2. Boundary value problems. 3. Differential equations, Elliptic—Numerical solutions. I. Mitrea, Marius. II. Taylor, Michael. III. Title. IV. Series.

QA3 .A57 no. 713
[QA649]
510 s—dc21
[516.3′73] 00-053585

Memoirs of the American Mathematical Society

This journal is devoted entirely to research in pure and applied mathematics.

Subscription information. The 2001 subscription begins with volume 149 and consists of six mailings, each containing one or more numbers. Subscription prices for 2001 are $494 list, $395 institutional member. A late charge of 10% of the subscription price will be imposed on orders received from nonmembers after January 1 of the subscription year. Subscribers outside the United States and India must pay a postage surcharge of $31; subscribers in India must pay a postage surcharge of $43. Expedited delivery to destinations in North America $35; elsewhere $130. Each number may be ordered separately; *please specify number* when ordering an individual number. For prices and titles of recently released numbers, see the New Publications sections of the *Notices of the American Mathematical Society.*

Back number information. For back issues see the *AMS Catalog of Publications.*

Subscriptions and orders should be addressed to the American Mathematical Society, P. O. Box 845904, Boston, MA 02284-5904. *All orders must be accompanied by payment.* Other correspondence should be addressed to Box 6248, Providence, RI 02940-6248.

Memoirs of the American Mathematical Society is published bimonthly (each volume consisting usually of more than one number) by the American Mathematical Society at 201 Charles Street, Providence, RI 02904-2294. Periodicals postage paid at Providence, RI. Postmaster: Send address changes to Memoirs, American Mathematical Society, P. O. Box 6248, Providence, RI 02940-6248.

This publication is indexed in *Science Citation Index*®, *SciSearch*®, *Research Alert*®, *CompuMath Citation Index*®, *Current Contents*®*/Physical, Chemical & Earth Sciences.*
Printed in the United States of America.

♾ The paper used in this book is acid-free and falls within the guidelines established to ensure permanence and durability.
Visit the AMS home page at URL: `http://www.ams.org/`

10 9 8 7 6 5 4 3 2 1 06 05 04 03 02 01

Contents

Abstract

The general aim of the present monograph is to study boundary-value problems for second-order elliptic operators in Lipschitz subdomains of Riemannian manifolds.

In the first part (§§1-4), we develop a theory for Cauchy type operators on Lipschitz submanifolds of codimension one (focused on boundedness properties and jump relations) and solve the L^p-Dirichlet problem, with p close to 2, for general second-order strongly elliptic systems. The solution is represented in the form of layer potentials and optimal nontangential maximal function estimates are established. This analysis is carried out under smoothness assumptions (for the coefficients of the operator, metric tensor and the underlying domain) which are in the nature of best possible.

In the second part of the monograph, §§5-13, we further specialize this discussion to the case of Hodge Laplacian $\Delta := -d\delta - \delta d$. This time, the goal is to identify all (pairs of) natural boundary conditions of Neumann type. Owing to the structural richness of the higher degree case we are considering, the theory developed here encompasses in a unitary fashion many basic PDE's of mathematical physics. Its scope extends to also cover Maxwell's equations, dealt with separately in §14.

The main tools are those of PDE's and harmonic analysis, occasionally supplemented with some basic facts from algebraic topology and differential geometry.

2000 *Mathematics Subject Classification.* Primary 35J55, 42B20, 58J05, 58J32, 58A14; Secondary 31B10, 31C12, 45E05, 78A30.

Key words and phrases. Elliptic systems, layer potentials, Dirichlet problem, Lipschitz domains, Hodge Laplacian, Maxwell's equations, Rellich estimates, harmonic fields.

Dedicated to the memory of Eugene Fabes

Introduction

We start with a brief explanation of the genesis of our results and then proceed with their systematic presentation.

In his pioneering work in the late 1950's and early 1960's A. P. Calderón was able to use symbolic calculus for singular integral operators in order to obtain many new results in Partial Differential Equations. In particular, he studied boundary problems for elliptic PDE in a smooth domain Ω, reducing them to pseudodifferential equations on $\partial\Omega$ via what is now called the Calderón projector. By way of contrast, when $\partial\Omega$ is merely C^1 or Lipschitzian, the corresponding boundary operator may fail to be pseudodifferential and, typically, can only be described in terms of singular integrals. This drastically affects the nature of the problem at hand, which becomes fundamentally harder.

Calderón's intuition in the 1960's that such problems are ultimately tractable via harmonic analysis techniques stimulated a long-term program with profound implications in the field. Some of the most striking advances were the proof of the L^2 boundedness of the Cauchy operator on C^1 and Lipschitz curves (A. P. Calderón [Ca1], R. Coifman, A. McIntosh and Y. Meyer [CMM]) and the treatment of the Laplacian with Dirichlet and Neumann boundary conditions on C^1 and Lipschitz domains in $\mathbb{R}^N$ by B. Dahlberg, E. Fabes, D. Jerison, C. Kenig, G. Verchota and their collaborators (cf. [Da], [FJR], [JK1], [Ve1], [DK]). An excellent account of these developments as well as a broader survey of the state of the art in this area up to the early 1990's can be found in Kenig's book [Ke2].

Nearly two decades ago, in his ICM plenary address ([Ca2]) A. P. Calderón also predicted that the method of integral equations on the boundary could be used successfully to deal with much more general elliptic systems than the Laplacian on Lipschitz domains in the Euclidean space. Subsequent investigations by E. Fabes and his circle of collaborators in the 1980's supported this conjecture but were limited to the case of homogeneous, constant-coefficient operators in the flat, Euclidean setting (cf. [Fa], [DKV], [FKV], [Ga]). Since the end of the 1980's, very few new elliptic boundary value problems have been attacked from this point of view (cf. [Mi3]). One notable related development was the treatment, in 1995, of the homogeneous higher order case ([PV]); however, this was done in the same analytical and geometrical context as before. Parenthetically, let us also note that the approach in [PV] relies on *a priori* dilation-free estimates for known solutions to such PDE's in smooth domains rather than the boundary methods alluded to above.

Quite recently, in [MT1], some new techniques were introduced that extended the study initiated in [FJR] and [Ve1] to include the Laplace-Beltrami operator on manifolds, in the scalar case. The main aim of the present monograph is to study general systems of second-order, strongly elliptic PDE in the global setting,

Received by the editor February 8, 1998

The first author was partially supported by UM Research Board and a UMC Summer Research Fellowship grant

The second author was partially supported by a UM Research Board grant and the NSF grant DMS-9870018

The third author was partially supported by NSF grant DMS-9600065 with an emphasis on the natural boundary problems for the Hodge Laplacian, on a Riemannian manifold with Lipschitz boundary, and with minimal smoothness requirements on its metric tensor. This requires both a considerable sharpening of the key techniques of [MT1] as well as the introduction of several important new ideas.

In the first part of the monograph (§§ 1–3), we develop a general theory of the Dirichlet problem for variable-coefficient, second-order, strongly elliptic systems of PDE's. In addition to working on a domain with only Lipschitz boundary, we place minimal hypotheses on the coefficients of the differential operators. Our main result in this part of the monograph (Theorem 3.1) deals with the global Dirichlet boundary problem for a general second-order, strongly elliptic, formally self-adjoint operator L, acting on sections of a vector bundle $\mathcal{E}$ over a compact, Riemannian manifold $\mathcal{M}$. It is assumed that the metric structures on $\mathcal{M}$ and $\mathcal{E}$ have C^1 coefficients and that, in local coordinates $U \subset \mathcal{M}$ over which $\mathcal{E}$ trivializes,

$$(Lu)^\alpha = \sum_{j,k}\sum_{\beta} \frac{\partial}{\partial x_j} a_{jk}^{\alpha\beta} \frac{\partial u^\beta}{\partial x_k} + \sum_{j}\sum_{\beta} b_j^{\alpha\beta} \frac{\partial u^\beta}{\partial x_j} + \sum_{\beta} d^{\alpha\beta} u^\beta \tag{0.1}$$

where, for some $\gamma > 0$ and $r > \dim \mathcal{M}$,

$$a_{jk}^{\alpha\beta} \in C^{1+\gamma}, \quad b_j^{\alpha\beta} \in H^{1,r}, \quad d^{\alpha\beta} \in L^r. \tag{0.2}$$

If, in addition, L satisfies a non-singularity hypothesis of the form

$$\forall\, D(\subseteq \mathcal{M}) \text{ Lipschitz domain,} \quad u \in H_0^{1,2}(D,\mathcal{E}),\ Lu = 0 \Longrightarrow u = 0 \text{ in } D, \tag{0.3}$$

(which is in fact necessary in the context of our result), then for any Lipschitz subdomain Ω of $\mathcal{M}$ the Dirichlet problem (in Dahlberg's sense)

$$\begin{cases} u \in C^0(\Omega,\mathcal{E}), \\ Lu = 0 \text{ in } \Omega, \\ u\big|_{\partial\Omega} = f \in L^2(\partial\Omega,\mathcal{E}), \end{cases} \tag{0.4}$$

has a unique solution satisfying the nontangential maximal function estimate

$$\|\mathcal{N}u\|_{L^2(\partial\Omega)} \le C\|f\|_{L^2(\partial\Omega,\mathcal{E})}. \tag{0.5}$$

In (0.4), the boundary trace is taken in a pointwise (nontangential limit) sense; precise definitions are given below.

An appropriate regularity result is valid when $f \in H^{1,2}(\partial\Omega,\mathcal{E})$ and global interior Sobolev estimates are established in each case. Moreover, analogous results are valid with the exponent 2 in (0.4)–(0.5) replaced by $p \in (2-\varepsilon, 2+\varepsilon)$, for some $\varepsilon = \varepsilon(\Omega,\mathcal{M},L) > 0$. Also, the formal self-adjointness hypothesis can be relaxed to

$$\text{degree}\,(L - L^*) \le 1 \tag{0.6}$$

which is optimal in the light of recent counterexamples in [VV] (they are originally given in the Euclidean context but can be lifted on a torus in a straightforward fashion). There it has been shown that uniqueness for the Dirichlet problem (0.4)–(0.5) may fail in Lipschitz domains for homogeneous, constant-coefficient, second-order systems satisfying the Legendre-Hadamard ellipticity condition but which are not symmetric. It is important to point out that, as a simple application of Plancherel's theorem shows, such operators always satisfy the non-singularity hypothesis (0.3); this is also an indication of the rather subtle interplay between the "variational" Dirichlet problem and the Dirichlet problem in Dahlberg's sense.

The strong ellipticity assumption which we make on L is also natural. In this vein, recall the well-known counterexample (due to A. V. Bitsadze [Bi]) which shows that mere ellipticity for second-order systems may not yield Fredholm-solvable problems in the context of Dirichlet boundary conditions. See also the discussion in [Mo2].

The result just described may be regarded as a global, higher-degree version of the "Dirichlet problem in non-smooth domains" theme in [JK2] where the authors deal with scalar-valued, divergence-form, variable (C^∞) coefficient, second-order elliptic operators in non-smooth Euclidean domains. A basic example of an operator L satisfying the hypotheses (0.1)–(0.3) above is provided by the Hodge Laplacian

$$\Delta = -\delta d - d\delta \tag{0.7}$$

acting on sections of the ℓ-th exterior power of the tangent bundle of a Riemannian manifold $\mathcal{M}$ whose metric tensor satisfies

$$g \in H^{2,r}(\mathcal{M}, \mathrm{Hom}(T\mathcal{M} \otimes T\mathcal{M}, \mathbb{R})), \quad \text{for some } r > \dim \mathcal{M}, \tag{0.8}$$

and in the absence of global ℓ-harmonic fields (i.e., differential forms of degree ℓ annihilated both by d and δ), $1 \le \ell \le \dim \mathcal{M} - 1$. Recall that, locally,

$$\Delta = \sum_{j,k} g^{jk} \frac{\partial^2}{\partial x_j \partial x_k} + \sum_j P_j(g) \frac{\partial}{\partial x_j} + Q(g) \tag{0.9}$$

where $P_j(g)$ and $Q(g)$ are matrix-valued functions with entries depending polynomially on $\partial^\alpha g_{jk}$, $\partial^\alpha g^{jk}$ for, respectively, $|\alpha| \le 1$ and $|\alpha| \le 2$.

The reader may well wonder about the nature of our smoothness assumptions in (0.1) and (0.3). In this connection, we point out that, contrary to the scalar-valued case when the De Giorgi-Nash-Moser theory applies, solutions to general elliptic systems with bounded, measurable coefficients may not even be in L^∞_{loc}. See [Gi] for counterexamples. In fact, even in the scalar-valued case, Caffarelli, Fabes and Kenig ([CFK]) have produced examples which show that, in the flat, Euclidean setting, the elliptic measure associated with $L = \sum_{j,k} \partial_j a_{jk} \partial_k$, where $a_{jk} \in L^\infty$, and the surface measure on $\partial\Omega$ may be mutually singular. As is well known, (0.6) fails in situations like these.

In the global, invariant setting, the case of Dirichlet boundary condition stands in sharp contrast to that of Neumann type in view of the absence of a global, intrinsic conormal derivative operator on $\partial\Omega$ associated with L (when no particular algebraic form is assumed). As such, the problem is connected with the validity of Green's formula for systems. In fact, as simple examples show (cf., e.g., [Ve2]) Neumann type problems need not be normally solvable; a very natural example in this regard is also offered by our Theorem 5.2. This raises the issue of investigating the "natural" boundary conditions for a given operator L.

In the second part of this monograph we take up the task of studying natural boundary conditions for the Hodge Laplacian (0.8) on a Riemannian manifold $\mathcal{M}$ in a systematic fashion. Our basic smoothness assumption remains (0.1); note that this is in the nature of best possible as far as making pointwise a.e. sense of (0.9) in a straightforward way is concerned. When $\ell = 0$, the formula (0.9) takes a simpler form and this allows for the relaxation $g \in C^1$. In this context, the corresponding scalar Dirichlet and Neumann boundary problems for the Laplace-Beltrami operator

$$\Delta u = \frac{1}{\sqrt{g}} \sum_{j,k} \frac{\partial}{\partial x_j} \left[g^{jk} \sqrt{g} \frac{\partial u}{\partial x_k} \right] \tag{0.10}$$

on Lipschitz subdomains of Riemannian manifolds have been treated in [MT1]. However, the structural richness of the higher degree case allows for a much larger variety of natural boundary value problems for harmonic forms which we formulate and solve on arbitrary Lipschitz domains $\Omega \subseteq \mathcal{M}$. The actual statements of these problems are too involved to be recorded here so, for the purpose of this introduction, we shall only single out pairs of boundary conditions that can be naturally prescribed in $L^2(\partial\Omega)$ for an ℓ-form u harmonic in Ω:

$$\begin{array}{llll}
\begin{cases} \nu \vee u, \\ \nu \vee du, \end{cases} & \begin{cases} \nu \wedge u, \\ \nu \wedge \delta u, \end{cases} & \begin{cases} \nu \vee \delta u, \\ \nu \vee du, \end{cases} & \begin{cases} \nu \wedge du, \\ \nu \wedge \delta u, \end{cases} \\
\begin{cases} \nu \vee du, \\ \nu \wedge \delta u, \end{cases} & \begin{cases} \nu \vee \delta u, \\ \nu \wedge du, \end{cases} & \begin{cases} \nu \vee u, \\ \nu \wedge du, \end{cases} & \begin{cases} \nu \wedge u, \\ \nu \vee \delta u, \end{cases} \\
\begin{cases} \nu \wedge du, \\ \nu \wedge \delta u, \end{cases} & \begin{cases} \nu \vee du, \\ \nu \vee \delta u, \end{cases} & \begin{cases} \nu \wedge u, \\ \nu \vee du, \end{cases} & \begin{cases} \nu \vee u, \\ \nu \wedge \delta u. \end{cases}
\end{array} \tag{0.11}$$

Here $\nu \in L^\infty(\partial\Omega, \Lambda^1 T\mathcal{M})$ is the outward unit conormal to $\partial\Omega$ and $\wedge$, $\vee$ stand, respectively, for the exterior and interior product of forms. Let us point out that these are *all* the (natural) combinations between the tangential and the normal components of u, du and δu on $\partial\Omega$.

As is explained later, by appropriately specializing the degree ℓ of the harmonic form u and/or the prescribed boundary values in (0.11), one obtains a variety of basic PDE's of mathematical physics in the non-smooth context (arising in such areas as hydrodynamics and electro-magnetism). In fact, the case of Maxwell's equations on manifolds, dealt with separately in § 14, is discussed in detail (this extends the Euclidean treatment in [Mi2] and [JM]). For all these, we develop a unified, very effective approach and many of our results are new even in the smooth setting; see § 5 for a more detailed discussion on this and related issues.

Another perspective from which the results in the second part of our monograph (§§ 5–13) can be understood has to do with the direction initiated by Hodge's work on harmonic integrals in the 1930's. The latter originated in Hodge's generalization to compact Riemannian manifolds of the potential theory used by Riemann in his study of Riemannian surfaces. Subsequent developments, due to Hodge himself and his immediate successors, de Rham, Kodaira, Kohn, Spencer, Duff, Friedrichs, Morrey (to cite just a few) have emphasized even more the basic role played in this theory by the natural generalizations of the classical Dirichlet and Neumann problems to this setting. Our results provide a bridge between this fundamental direction in potential theory and the line of work emerging from Calderón's program alluded to before. The main question that we address (and thoroughly answer) is that of the effectiveness of the method of layer potentials in the global, higher degree context, arbitrary topology and in the presence of nonsmooth structures (a rough metric tensor, Lipschitz boundaries and discontinuous boundary data).

The organization of the monograph is as follows. The first chapter contains an extension of the Calderón-Zygmund theory (as presented in, e.g., [Me]) to the vector bundle-valued case. Here we work with singular integral operators whose kernels are the Schwartz kernels of classical pseudodifferential operators in $\mathrm{OP}C^0S^{-1}_{\mathrm{cl}}(\mathcal{E},\mathcal{E})$. In § 2 we analyze the Schwartz kernels of L^{-1} and of PL^{-1}; the differential operator L, introduced in (0.2), acts here between appropriate Sobolev spaces and P is a first-order differential operator on $\mathcal{M}$. We show that the analysis of § 1 applies to the most singular part of PL^{-1}, and also establish appropriate estimates on the less singular remainder.

General second-order, strongly elliptic systems are discussed in § 3 where problems of the type (0.5)–(0.6) are treated. Besides the machinery developed in §§ 1 − 2, another key ingredient in the proof of Theorem 3.1 (the main result of this chapter, outlined before) is a variable-coefficient Rellich-type identity due to [Ne], [PW] and others. A fundamental case for which the general theory of § 3 applies is that of the Hodge Laplacian Δ, separately discussed in § 4 (along with other related operators). Actually, in order to avoid topological complications, we find it convenient to work with a Schrödinger operator $\Delta - V$, where the positive, scalar-valued potential V vanishes identically on $\bar{\Omega}$. Here we rely on a deep unique continuation theorem of Aronszajn, Krzywicki and Szarski ([AKS]) for differential forms on Riemannian manifolds equipped with a Lipschitz metric tensor.

Natural boundary value problems for the Hodge Laplacian on Lipschitz subdomains of Riemannian manifolds are formulated in § 5, and essentially the remaining part of the monograph is devoted to presenting the proofs of the results stated there. In a first stage, relevant boundary layer potential operators are devised in § 6 and the next three consecutive chapters are concerned with proving Fredholm properties for them. This is a complex task for which we use a broad repertoire of techniques from such distinct areas as harmonic analysis, algebraic topology and differential geometry. In particular, in § 7 we prove some new Rellich-type identities for arbitrary harmonic differential forms on Lipschitz domains. Unlike the case of scalar-valued functions or vector fields however, these yield only a partial result as far as Fredholmness is concerned (cf. § 8). We augment this analysis with certain duality and localization arguments, presented in §§ 9 − 10, which eventually allow us to conclude the desired result in full strength.

The topology of the underlying domain plays a major role in the problem under discussion. For instance, there are natural topological obstructions to the unique solvability of our boundary problems as well as to the invertibility of the boundary operators under discussion. Such connections, involving certain singular homology groups of the domain, are brought forward in § 11 (and Appendix B). Among the main ingredients used here, we note a regularity result, allowing for the transition between variational methods in Ω and boundary integral techniques (Theorem 11.2), and de Rham-Hodge theory, dealing with the topological information encoded in the problem. Finally, the proofs of the results announced in § 5 are completed in §§ 12 − 13, while § 14 contains an application to boundary value problems for the time-harmonic Maxwell equations.

Acknowledgments. The methods and results of this monograph owe a great deal to earlier work of many people. On a personal note, the first two authors would like to take this opportunity to acknowledge the deep, lasting influence that Eugene B. Fabes has had on their work. Thanks are also due to Alan McIntosh, Taduesz Iwaniec and Steve Hofmann for several enlightening discussions in the last four years when many of the ideas presented here eventually crystallized.

CHAPTER 1

Singular Integrals on Lipschitz Submanifolds of Codimension One

Let $\mathcal{M}$ be a smooth, connected, compact, boundaryless manifold of real dimension m, and denote by $T\mathcal{M}$, $T^*\mathcal{M}$ its tangent and cotangent bundles, respectively. Although many of our results easily extend to the non-orientable case, in order to facilitate the presentation we shall consider $\mathcal{M}$ oriented. We also assume that $\mathcal{M}$ is equipped with a Lipschitzian Riemannian metric tensor $g = \sum_{j,k} g_{jk} dx_j \otimes dx_k$, $g_{jk} \in \mathrm{Lip}\,(\mathcal{M})$. As is customary, $\det\,(g_{jk})$ is also denoted by g so that, if $d\mathrm{Vol}$ stands for the corresponding volume form, then $d\mathrm{Vol} = \sqrt{g(x)}dx_1 \wedge \cdots \wedge dx_m$ locally. We also set $(g^{jk})_{j,k} := ((g_{jk})_{j,k})^{-1}$.

Consider now two smooth vector bundles $\mathcal{E}, \mathcal{F} \to \mathcal{M}$ endowed with the Lipschitzian Hermitian structures $\langle\cdot,\cdot\rangle_{\mathcal{E},x}$ and $\langle\cdot,\cdot\rangle_{\mathcal{F},x}$, respectively (in the sequel, we shall omit the subscripts $\mathcal{E}, \mathcal{F}$ whenever obvious from the context). Each time we shall agree that $L^1(\mathcal{M},\mathcal{E}) \hookrightarrow \mathcal{D}'(\mathcal{M},\mathcal{E})$, the embedding of integrable sections into the space of distributions with coefficients in $\mathcal{E}$ is made in such a way that any integrable section f is identified with the functional

$$C^\infty(\mathcal{M},\mathcal{E}) \ni \phi \mapsto \iint_{\mathcal{M}} \langle f(x), \phi(x)\rangle_x \, d\mathrm{Vol}(x). \tag{1.1}$$

We shall find it necessary to work with classes of symbols $S^\ell_{1,0}$ which only exhibit a limited amount of regularity in the spatial variable (while still C^∞ in the Fourier variable). The generic case in the scalar-valued Euclidean setting is as follows. For a normed function space $(\mathcal{X}, \|\cdot\|_{\mathcal{X}})$ such that $C^\infty_{\mathrm{comp}} \subseteq \mathcal{X} \subseteq C^0$ (most typically $\mathcal{X} = C^\mu$ or a similar smoothness space) we say

$$q(x,\xi) \in \mathcal{X}S^\ell_{1,0} \Longleftrightarrow \|D^\alpha_\xi q(\cdot,\xi)\|_{\mathcal{X}} \le C_\alpha (1+|\xi|)^{\ell-|\alpha|}, \quad \forall \alpha \ge 0. \tag{1.2}$$

The class of pseudodifferential operators $Q(x,D)$ with symbols $q(x,\xi) \in \mathcal{X}S^\ell_{1,0}$ will be denoted by $\mathrm{OP}\mathcal{X}S^\ell_{1,0}$. As usual, we write $\mathrm{OP}\mathcal{X}S^\ell_{\mathrm{cl}}$ for the subclass of *classical* pseudodifferential operators, i.e. $Q(x,D) \in \mathrm{OP}\mathcal{X}S^\ell_{1,0}$ with symbol $q(x,\xi) \sim q_\ell(x,\xi) + q_{\ell-1}(x,\xi) + \cdots$, where $q_j(x,\xi)$ has $\mathcal{X}$-order regularity in x and is smooth and homogeneous of degree j in ξ for $|\xi| \ge 1$, $j = \ell, \ell-1, \ldots,$.

Next, we denote by $\mathrm{OP}\mathcal{X}S^\ell_{1,0}(\mathcal{E},\mathcal{F})$, etc, the collection of pseudodifferential operators mapping sections of the vector bundle $\mathcal{E}$ into sections of the vector bundle $\mathcal{F}$ which, in local coordinates of $\mathcal{M}$ and over local trivializations of $\mathcal{E}$, $\mathcal{F}$, can be represented as matrices with entries from $\mathrm{OP}\mathcal{X}S^\ell_{1,0}$. Finally, $\emptyset\mathrm{P}\mathcal{X}S^\ell_{\mathrm{cl}}(\mathcal{E},\mathcal{F})$ will consist of all formal adjoints of operators in $\mathrm{OP}\mathcal{X}S^\ell_{\mathrm{cl}}(\mathcal{F},\mathcal{E})$.

Going further, for $Q(x,D) \in \mathrm{OP}C^0S^{-1}_{\mathrm{cl}}(\mathcal{E},\mathcal{F})$ a classical pseudodifferential operator, we set

$$k(Q) \in \mathcal{D}'(\mathcal{M}\times\mathcal{M},\, \mathcal{F}\otimes\mathcal{E}) \tag{1.3}$$

for its Schwartz kernel, and let

$$\sigma(Q) : T^*\mathcal{M}\setminus 0 \to \mathrm{Hom}\,(\mathcal{E},\mathcal{F}) \tag{1.4}$$

stand for its principal symbol, $\sigma(Q) \in C^0S^{-1}_{\mathrm{cl}}(\mathcal{E},\mathcal{F})$.

Fix an arbitrary Lipschitz domain $\Omega \subseteq \mathcal{M}$ (see, e.g., [Mo3], [MT1] for a definition) and denote by $\nu(x) \in T^*_x\mathcal{M}$ the outward unit conormal (well defined) at almost every $x \in \partial\Omega$. Also, $d\sigma$ is the surface measure induced by g on $\partial\Omega$, where the latter is regarded as a Lipschitz submanifold of codimension one in $\mathcal{M}$. We shall work with "natural" spaces like $L^p(\partial\Omega,\mathcal{E})$, $H^{1,p}(\partial\Omega,\mathcal{E})$, etc.

Next, we introduce layer potential operators by formally writing, for $f : \partial\Omega \to \mathcal{E}$,

$$\begin{aligned} K(Q)f(x) &:= \mathrm{p.v.}\int_{\partial\Omega} \langle k(Q)(x,y), f(y)\rangle_y \, d\sigma(y) \\ &= \lim_{\varepsilon\to 0} \int_{y\in\partial\Omega,\, r(x,y)>\varepsilon} \langle k(Q)(x,y), f(y)\rangle_y \, d\sigma(y), \qquad x \in \partial\Omega, \end{aligned} \tag{1.5}$$

where $r(x,y)$ stands for the geodesic distance between x and $y \in \mathcal{M}$, with respect to some smooth background metric g_0 (perhaps different from g). Also, we set

$$\mathcal{K}(Q)f(x) := \int_{\partial\Omega} \langle k(Q)(x,y), f(y)\rangle_y \, d\sigma(y), \qquad x \in \mathcal{M}\setminus\partial\Omega. \tag{1.6}$$

Similar considerations apply to pseudodifferential operators in $\emptyset\mathrm{P}C^0S^{-1}_{\mathrm{cl}}(\mathcal{E},\mathcal{F})$.

To state the main result of this chapter we need some more notation. We set $\Omega_+ := \Omega$, $\Omega_- := \mathcal{M}\setminus\overline{\Omega}$, and let $\cdot|_{\partial\Omega_\pm}$ be the nontangential boundary trace operators on $\partial\Omega_\pm$. That is,

$$u|_{\partial\Omega_\pm}(x) := \lim_{y\in\gamma_\pm(x)} u(y), \quad x \in \partial\Omega,$$

where $\gamma_\pm(x) \subseteq \Omega_\pm$ are appropriate nontangential approach regions. Finally, $\mathcal{N}$ stands for the nontangential maximal operator defined for sections $u : \Omega_\pm \to \mathcal{E}$ by

$$\mathcal{N}u(x) := \sup_{y\in\gamma_\pm(x)} |u(y)|, \quad x \in \partial\Omega.$$

THEOREM 1.1. *Let $\mathcal{E},\mathcal{F} \to \mathcal{M}$ be as above and let $Q(x,D) \in OPC^0S^{-1}_{\mathrm{cl}}(\mathcal{E},\mathcal{F})$ be such that $\sigma(Q)(x,\xi)$ is odd in $\xi \in T^*_x\mathcal{M}$. Then, for each $f \in L^p(\partial\Omega,\mathcal{E})$, with $1 < p < \infty$, $K(Q)f(x)$ exists pointwise at almost every boundary point $x \in \partial\Omega$ and*

$$K(Q) : L^p(\partial\Omega,\mathcal{E}) \to L^p(\partial\Omega,\mathcal{F}) \tag{1.7}$$

is a bounded operator. Also, there exists $C > 0$ so that

$$\|\mathcal{N}(\mathcal{K}(Q)f)\|_{L^p(\partial\Omega)} \le C\|f\|_{L^p(\partial\Omega,\mathcal{E})}, \qquad \forall f \in L^p(\partial\Omega,\mathcal{E}), \tag{1.8}$$

and $\mathcal{K}(Q)f$ has a nontangential boundary trace at almost every boundary point. More specifically,

$$\mathcal{K}(Q)f|_{\partial\Omega_\pm} = \mp\tfrac{1}{2}i\,\sigma(Q)(\cdot,\nu)f + K(Q)f \quad \text{a.e. on } \partial\Omega. \tag{1.9}$$

Furthermore, if actually $Q(x,D) \in OPC^r S^{-1}_{\mathrm{cl}}(\mathcal{E},\mathcal{F})$ for some $r > 1/2$, then

$$\mathcal{K}(Q) : L^2(\partial\Omega,\mathcal{E}) \to H^{1/2,2}(\Omega,\mathcal{F}) \tag{1.10}$$

is a bounded operator.

Finally, similar results are valid for pseudodifferential operators in the space $\emptyset PC^0 S^{-1}_{\mathrm{cl}}(\mathcal{E},\mathcal{F})$. Moreover, in this latter case, (1.10) is valid without any extra smoothness assumption.

PROOF. The problem localizes and, given the invariance of the class of pseudodifferential operators and symbols under discussion, it can be transported in $\mathbb{R}^m$ via coordinate mappings. By fixing local orthonormal frames in $\mathcal{E}$ and $\mathcal{F}$, $Q(x,D)$ can be canonically identified (locally) with a matrix of classical pseudodifferential operators $(Q_{jk}(x,D))_{j,k}$, while $k(Q)(x,y)$ can be identified with a matrix-valued distribution $(b_{jk}(x,y))_{j,k}$. The latter is such that the Schwartz kernel of $Q_{jk}(x,D)$ with respect to the Euclidean volume form is $b_{jk}(x,y)\sqrt{g(y)}$. Going further, $\partial\Omega$ becomes a Lipschitz hypersurface Γ in $\mathbb{R}^m$ and the pull-back of $d\sigma$ to Γ is

$$\sqrt{g}\left(\sum_{j,k} g^{jk} n_j n_k\right)^{1/2} d\sigma_\Gamma \tag{1.11}$$

where n is the unit conormal to Γ (with respect to the Euclidean metric) and $d\sigma_\Gamma$ stands for the canonical surface measure on Γ (inherited from the Euclidean structure). Also, the unit conormal to $\partial\Omega$ becomes

$$\nu = \left(\sum_{j,k} g^{jk} n_j n_k\right)^{-1/2} n. \tag{1.12}$$

Consequently, $\mathcal{K}(Q)$ can be identified with the operator-valued matrix $(\mathcal{B}_{jk})_{j,k}$ where

$$\mathcal{B}_{jk} f(x) := \int_\Gamma b_{jk}(x,y)\sqrt{g(y)}\left(\sum_{j,k} g^{jk}(y) n_j(y) n_k(y)\right)^{1/2} f(y)\, d\sigma_\Gamma(y). \tag{1.13}$$

Let us consider the jump formula (1.9). With the above notation, Proposition 1.9 in [MT1] gives that

(1.14)

$$\begin{aligned}
&\mathcal{B}_{jk} f\big|_{\Gamma_\pm}(x) \\
&\quad = \mp \tfrac{1}{2} i\sigma(Q_{jk})(x,n(x))\left(\sum_{j,k} g^{jk}(x) n_j(x) n_k(x)\right)^{1/2} f(x) \\
&\qquad + \lim_{\varepsilon\to 0} \int_{y\in\Gamma,\, r(x,y)>\varepsilon} b_{jk}(x,y)\sqrt{g(y)}\left(\sum_{j,k} g^{jk}(y) n_j(y) n_k(y)\right)^{1/2} f(y)\, d\sigma_\Gamma(y) \\
&\quad = \mp \tfrac{1}{2} i\sigma(Q_{jk})(x,\nu(x)) f(x) + K_{jk} f(x).
\end{aligned}$$

where we have used the fact that $\sigma(Q_{jk})(x,\xi)$ is homogeneous of degree -1 in ξ, and we have denoted by K_{jk} the operator given by the principal-value integral in (1.14).

Since $K(Q) \equiv (K_{jk})_{j,k}$, the analysis above amounts to

$$\begin{aligned}\mathcal{K}(Q)f\big|_{\partial\Omega_\pm}(x) &= \mp\tfrac{1}{2}i\,\sigma(Q)(x,\nu(x))f(x)\\ &\quad+\lim_{\varepsilon\to 0}\int_{y\in\partial\Omega,\,r(x,y)>\varepsilon}\langle k(Q)(x,y),f(y)\rangle_y\,d\sigma(y)\end{aligned}\tag{1.15}$$

which is precisely (1.9). Let us remark that (1.7) and (1.8) follow in a similar fashion from the corresponding scalar-valued theory in [MT1] and we omit the details.

Finally, we turn our attention to (1.10). We shall make a series of reductions which we now proceed to explain. First, once again the problem localizes and, therefore, it suffices to show the following statement. For $\Omega \subseteq \mathbb{R}^m$ the (unbounded) Lipschitz domain above the graph of a Lipschitz function $\phi : \mathbb{R}^{m-1} \to \mathbb{R}$, the mapping

$$L^2_{\rm comp}(\partial\Omega) \ni f \mapsto u(x) := \int_{\partial\Omega} b(x, x-y)f(y)\,d\sigma(y) \in H^{1/2,2}_{\rm loc}(\overline{\Omega})\tag{1.16}$$

is bounded; here $b(x,z)$ is such that

$$\|\partial_z^\beta b(\cdot,z)\|_{C^r(\mathbb{R}^m)} \text{ is bounded on } S^{m-1} \text{ for any } \beta,\tag{1.17}$$

and

$$b(x,z) \text{ is odd and positive homogeneous of degree } -(m-1) \text{ in } z.\tag{1.18}$$

Second, expanding the function $b(x,z)$ in spherical harmonics via the classical method of Calderón and Zygmund, as in the proof of Proposition 1.2 in [MT1] (used to prove (1.7)–(1.8) here), it is enough to further treat the case when $b(x,z)$ is *independent of* x. Assume that this is the case and write $b(z) := b(x,z)$, $u(x) := \int_{\partial\Omega} b(x-y)f(y)\,d\sigma(y)$, $x \in \Omega$. Now, by the implication "$(b) \Longrightarrow (a)$" of Theorem 4.1 in [JK3] (this is stated there for harmonic functions but the argument is of a purely real-variable nature), the mapping property (1.10) is a consequence of the square function estimate

$$\iint_\Omega |u|^2\,dx + \iint_\Omega |\nabla u(x)|^2 {\rm dist}\,(x,\partial\Omega)\,dx \le C\int_{\partial\Omega}|f|^2\,d\sigma,\tag{1.19}$$

which we claim is valid for some $C > 0$, independent of f. Indeed, bounding the first integral in the left-side is a consequence of (1.8) with $p = 2$. Going further, via the bi-Lipschitz change of variables $\mathbb{R}^m_+ \ni (t,x') \mapsto (\phi(x')+t,x') \in \Omega$, the corresponding estimate for the second solid integral above follows from the boundedness of

$$\begin{aligned}&T^j : L^2(\mathbb{R}^{m-1},dx') \to L^2(\mathbb{R}^m_+,\tfrac{dt}{t}dx'),\\ &T^j f(t,x') := \int_{\mathbb{R}^{m-1}} K^j_t(x',y')f(y')\,dy'\end{aligned}\tag{1.20}$$

for $j = 1,..,m$, where the family of kernels $\{K^j_t(x',y')\}_{t>0}$ is given by

$$K^j_t(x',y') := t\,(\partial_j b)(\phi(x')-\phi(y')+t,x'-y').\tag{1.21}$$

The approach we present utilizes ideas developed in [CJ] and [Ho]. First, based on (1.17) and (1.18), it is not difficult to check that the family $\{K_t^j(x',y')\}_{t>0}$ is standard, i.e., there hold

$$\begin{aligned} |K_t^j(x',y')| &\le C\,t(t+|x'-y'|)^{-m} \\ |\nabla_{x'}K_t^j(x',y')| + |\nabla_{y'}K_t^j(x',y')| &\le C\,t(t+|x'-y'|)^{-(m+1)}. \end{aligned} \tag{1.22}$$

Therefore, by the results of M. Christ and J.-L. Journé [CJ] (cf. also Theorem 20, pp. 69 in [Ch]), in order to check (1.20) it suffices to prove that

$$T^j(1)\,\tfrac{dt}{t}dx' \text{ is a Carleson measure in } \mathbb{R}^m_+. \tag{1.23}$$

To this end, fix a smooth, even, positive function $\theta : \mathbb{R}^{m-1} \to \mathbb{R}$ with $\int \theta(x')\,dx' = 1$ and, as usual, set $\theta_t(x') := t^{-(m-1)}\theta(x'/t)$. We write $T^j(1) = (T^j(1) - \tilde{T}^j(1)) + \tilde{T}^j(1)$ where

$$\tilde{T}^j f(t,x') := \int_{\mathbb{R}^{m-1}} \tilde{K}_t^j(x',y')f(y')\,dy' \tag{1.24}$$

with

$$\tilde{K}_t^j(x',y') := t\,(\partial_j b)(\langle\nabla_{x'}(\theta_t * \phi)(x'),(x'-y')\rangle + t, x'-y'). \tag{1.25}$$

That $(T^j - \tilde{T}^j)(1)$ is the density of a Carleson measure can be proved by an adaptation of the techniques in [Ho]. For the convenience of the reader, we give a brief outline of the main steps. Fix x_0' in $\mathbb{R}^{m-1}$, $r>0$, and split

$$(T^j - \tilde{T}^j)(1) = (T^j - \tilde{T}^j)(\chi_{B(x_0',100r)}) + (T^j - \tilde{T}^j)(\chi_{\mathbb{R}^{m-1}\setminus B(x_0',100r)}),$$

where, for any set A, χ_A stands for the characteristic function of A. Now, a crude estimate based on (1.22) and a similar set of estimates for $\tilde{K}_t^j$ gives that

$$\int_0^r \int_{B(x_0',r)} |(T^j - \tilde{T}^j)(\chi_{B(x_0',100r)})(t,x')|^2\,dx'\,\tfrac{dt}{t} \le C\,r^{m-1} \tag{1.26}$$

and we are left with proving something similar with $B(x_0',100r)$ replaced by its complement in $\mathbb{R}^{m-1}$. First, from (1.17)–(1.18) and the mean-value theorem we have

$$|K_t^j(x',y') - \tilde{K}_t^j(x',y')| \le C\,t(t+|x'-y'|)^{-(m+1)}|E_\phi(t,x',y')|, \tag{1.27}$$

where

$$E_\phi(t,x',y') := \phi(x') - \phi(y') - \langle\nabla_{x'}(\theta_t * \phi)(x'),(x'-y')\rangle. \tag{1.28}$$

This is used in conjunction with the following lemma (itself a direct consequence of Plancherel's theorem), implicit in the work of S. Hofmann:

LEMMA 1.2 [Ho]. *Assume that the Lipschitz function ϕ is compactly supported in $\mathbb{R}^{m-1}$. Then, for some positive constant $C = C(\theta,m)$,*

$$\int_0^\infty t^{-m-1} \int_{\mathbb{R}^{m-1}} \int_{|x'-y'|\le t} |E_\phi(t,x',y')|^2\,dy'\,dx'\,\tfrac{dt}{t} \le C\|\nabla\phi\|^2_{L^2(\mathbb{R}^{m-1})}. \tag{1.29}$$

Now, if $\Phi \in C^\infty(\mathbb{R})$, $0 \le \Phi \le 1$, is such that $\operatorname{supp}\Phi \subseteq [-150r,150r]$, $\Phi \equiv 1$ on $[-125r,125r]$, and if we set $\tilde{\phi}(x') := \Phi(|x'-x_0'|^2)(\phi(x')-\phi(x_0'))$, then the expression $(T^j - \tilde{T}^j)(\chi_{\mathbb{R}^{m-1}\setminus B(x_0',100r)})(t,x')$ does not change for $x' \in B(x_0',2r)$ if we replace

ϕ by $\tilde{\phi}$. Assuming that this is the case and noting that $\|\nabla\tilde{\phi}\|^2_{L^2(\mathbb{R}^{m-1})} \le Cr^{m-1}$, the desired estimate easily follows from (1.27), Schwarz's inequality, (1.29), plus a natural rescaling argument. We omit the details.

Finally, we claim that

$$\tilde{T}^j(1)(t,x') = \int_{\mathbb{R}^{m-1}} t\,(\partial_j b)(\langle\nabla_{x'}(\theta_t * \phi)(x'),(x'-y')\rangle + t, x'-y')\,dy' \equiv 0. \tag{1.30}$$

In fact, this is a particular case of the elementary lemma presented below; this completes the proof of (1.23) and, with it, the proof of Theorem 1.1. ■

LEMMA 1.3. *Let $F : \mathbb{R}^m \setminus \{0\} \to \mathbb{R}$ be an even function which is positive homogeneous of degree $-m$. Then for any $a \in \mathbb{R}^{m-1}$ and any $t > 0$ we have that*

$$\begin{aligned} \int_{\mathbb{R}^{m-1}} F(a\cdot y' + t, y')\,dy' &= \frac{1}{2t}\int_{S^{m-2}}\int_{-\infty}^{\infty} F(s,\omega)\,ds\,d\omega \\ &= \int_{\mathbb{R}^{m-1}} F(t,y')\,dy'. \end{aligned} \tag{1.31}$$

In particular, if F is some partial derivative of a function G, homogeneous of degree $-(m-1)$, i.e., $F = \partial_j G$, $j \in \{1,..,m\}$, then $\int_{\mathbb{R}^{m-1}} F(a\cdot y' + t, y')\,dy' = 0$ for any $a \in \mathbb{R}^{m-1}$, $t > 0$.

PROOF. Passing to polar coordinates $y' = r\omega$, $r > 0$, $\omega \in S^{m-2}$, and using the homogeneity of F we may write

$$\int_{\mathbb{R}^{m-1}} F(a\cdot y' + t, y')\,dy' = \int_{S^{m-2}}\int_0^{\infty} F(a\cdot\omega + t/r, \omega)\,r^{-2}dr\,d\omega. \tag{1.32}$$

Now, setting $s := a\cdot\omega + t/r$ this becomes $t^{-1}\int_{S^{m-2}}\int_{a\cdot\omega}^{\infty} F(s,\omega)\,ds\,d\omega$ so that, making $(s,\omega) \mapsto (-s,-\omega)$ and using the fact that F is even, readily leads to

$$\frac{1}{t}\int_{S^{m-2}}\int_{a\cdot\omega}^{\infty} F(s,\omega)\,ds\,d\omega = \frac{1}{2t}\int_{S^{m-2}}\int_{-\infty}^{\infty} F(s,\omega)\,ds\,d\omega. \tag{1.33}$$

This gives the first equality in (1.31). Furthermore, the integral in the right side of (1.33) is independent of $a \in \mathbb{R}^{m-1}$ and, hence, so is the original one. In particular, its value does not change if we take $a = 0$ and this is precisely what the second equality in (1.31) says.

Finally, if $F = \partial_j G$ with $j \ne 1$ then the third integral in (1.31) vanishes, whereas if $F = \partial_1 G$ the second one does so. ■

A typical context to which Theorem 1.1 applies is when $Q(x,D) \in \mathrm{OP}S^{-1}_{\mathrm{cl}}(\mathcal{E},\mathcal{F})$ has the form $P(x,D)L^{\#}(x,D)$ (or its formal adjoint). Here $P(x,D) \in \mathrm{Diff}_1(\mathcal{E},\mathcal{F})$ is a first order differential operator and $L^{\#}(x,D) \in \mathrm{OP}S^{-2}_{\mathrm{cl}}(\mathcal{E},\mathcal{F})$ is a parametrix for an elliptic operator $L(x,D) \in \mathrm{OP}S^{2}_{\mathrm{cl}}(\mathcal{F},\mathcal{E})$ with even principal symbol. In fact, similar considerations are valid under considerably more relaxed smoothness assumptions on the coefficients of $L(x,D)$. This will become apparent in the next chapter where a thorough investigation of the main singularity of the Schwartz kernel of $L(x,D)^{-1}$ is presented (under hypotheses guaranteeing the existence of this inverse).

CHAPTER 2

Estimates on Fundamental Solutions

Assume that $\mathcal{M}$ is a smooth compact manifold of real dimension m, equipped with a C^1 metric tensor. Also, let $\mathcal{E}$, $\mathcal{F} \to \mathcal{M}$ be smooth vector bundles, equipped with C^1 Hermitian structures and connections whose connection coefficients, with respect to some smooth local trivializations, are continuous.

Consider $L(x, D)$, a second-order differential operator mapping C^2 (global) sections in $\mathcal{E}$ into (global) measurable sections in $\mathcal{F}$. If $U \subseteq \mathcal{M}$ is open, χ is a coordinate map of U into $\mathbb{R}^m$, and ϕ is a scalar-valued measurable function with support in U, we introduce the operator ϕ_χ, mapping functions defined in U into functions defined in $\chi(U)$, by setting

$$\phi_\chi g(y) := \begin{cases} \phi(\chi^{-1}(y))g(\chi^{-1}(y)), & \text{for any } y \in \chi(U), \\ 0, & \text{otherwise.} \end{cases}$$

Similarly, we introduce

$$\phi^\chi f(x) := \begin{cases} \phi(x)f(\chi(x)), & \text{for } x \in U, \\ 0, & \text{otherwise.} \end{cases}$$

Now, if $\{\theta_\alpha\}_\alpha$ and $\{\gamma_\beta\}_\beta$ are bases of $\mathcal{E}$ and $\mathcal{F}$, respectively, over U, and if ϕ, ψ are arbitrary real-valued cutoff functions supported in U, then it is assumed that

$$\phi L \psi\Big(\sum_\alpha f_\alpha \theta_\alpha\Big) = \sum_{\alpha,\beta} (\phi^\chi L_{\alpha\beta} \psi_\chi f_\alpha)\gamma_\beta \tag{2.1}$$

where

$$L_{\alpha\beta} h := \sum_{j,k} \partial_j a_{jk}^{\alpha\beta}(y)\partial_k h + \sum_j b_j^{\alpha\beta}(y)\partial_j h + d^{\alpha\beta}(y)h. \tag{2.2}$$

Then, in local coordinates, the principal symbol $\sigma(L) : T^*\mathcal{M} \setminus 0 \to \mathrm{Hom}(\mathcal{E}, \mathcal{F})$ is given by

$$\sigma(L)(x, \xi)\Big(\sum_\alpha f_\alpha \theta_\alpha\Big) := -\sum_{\alpha,\beta}\sum_{j,k} a_{jk}^{\alpha\beta}(\chi(x))\xi_j \xi_k f_\alpha(x)\gamma_\beta(x) \tag{2.3}$$

for any $x \in U$, $\xi \in T_x^*\mathcal{M}$, $\xi = \sum_j \xi_j d\chi_j \neq 0$. It is well known that this definition is independent of the particular choice of coordinates. Recall that L is called *elliptic* if $\sigma(L)(x, \xi) : \mathcal{E}_x \to \mathcal{F}_x$ is invertible for each $x \in \mathcal{M}$ and $\xi \in T_x^*\mathcal{M} \setminus 0$.

Also, L is called *strongly elliptic* provided $\mathcal{E} = \mathcal{F}$ and there exists $C > 0$ such that

$$\langle -\sigma(L)(x, \xi)\eta, \eta\rangle_x \geq C\langle \eta, \eta\rangle_x, \tag{2.4}$$

for any $x \in \mathcal{M}$, $\xi \in T_x^*\mathcal{M}$ with $|\xi|_x = 1$, and $\eta \in \mathcal{E}_x$.

In this chapter, we aim to study the Schwartz kernel of L^{-1}, under appropriate hypotheses guaranteeing that L is invertible. We begin with a regularity result.

PROPOSITION 2.1. *Let $\mathcal{O}$ be an open subset of $\mathcal{M}$. Assume L is an elliptic operator given by (2.1), with coefficients satisfying*

$$a_{jk}^{\alpha\beta} \in C^1, \quad b_j^{\alpha\beta} \in L^\infty, \quad d^{\alpha\beta} \in L^r. \tag{2.5}$$

Also assume

$$1 < p \leq q < \infty, \quad -1 \leq \sigma < 0, \tag{2.6}$$

and

$$r > max\{q, \tfrac{m}{2}\}. \tag{2.7}$$

Then

$$u \in H^{1,p}_{\rm loc}(\mathcal{O},\mathcal{E}), \ Lu \in H^{\sigma,q}_{\rm loc}(\mathcal{O},\mathcal{F}) \Longrightarrow u \in H^{2+\sigma,q}_{\rm loc}(\mathcal{O},\mathcal{E}) \tag{2.8}$$

and natural estimates hold.

A family of examples of particular interest is given by the Hodge Laplacian Δ on differential forms, when the metric tensor belongs to $H^{2,r}$, for some $r > m = \dim \mathcal{M}$ or, more generally, $L = \Delta - V$ with $V \in L^r(\mathcal{M})$.

PROOF. For notational simplicity, we denote $H^{s,q}_{\rm loc}(\mathcal{O},\mathcal{E})$ by $H^{s,q}$, etc. Suppose that, in local coordinates, L has the form

$$Lu = \sum_{j,k} \partial_j A^{jk}(x)\partial_k u + \sum_j B^j(x)\partial_j u - V(x)u, \tag{2.9}$$

where $A^{jk} := \left(a^{\alpha\beta}_{jk}\right)_{\alpha,\beta}$, $B^j := \left(b^{\alpha\beta}_j\right)_{\alpha,\beta}$ and $V := -\left(d^{\alpha\beta}\right)_{\alpha,\beta}$ are matrix-valued functions with entries satisfying (2.5). We use a symbol decomposition on the first-order differential operator $A_j := \sum_k A^{jk}(x)\partial_k$, of a sort introduced by [KN] and [Bo] (cf. also the exposition in [Ta1]). Picking $\delta \in (0,1)$, write

$$A_j = A_j^{\#} + A_j^{b}, \quad \text{where } A_j^{\#} \in \mathrm{OP}S^1_{1,\delta} \text{ and } A_j^b \in \mathrm{OP}C^1 S^{1-\delta}_{1,\delta}. \tag{2.10}$$

Then set $L^{\#} := \sum \partial_j A_j^{\#} \in \mathrm{OP}S^2_{1,\delta}$, elliptic, and denote by $E^{\#} \in \mathrm{OP}S^{-2}_{1,\delta}$ a parametrix of $L^{\#}$. Make all pseudodifferential operators properly supported. We have, on $\mathcal{O}$,

$$u = E^{\#}f - \sum_j E^{\#}\partial_j A_j^b u - \sum_j E^{\#}B^j\partial_j u + E^{\#}Vu, \quad \text{mod } C^\infty, \tag{2.11}$$

where $f := Lu$. Note that the hypotheses imply $E^{\#}f \in H^{2+\sigma,q}$. Furthermore, by (a special case of the) results in [Bou] (cf. also Proposition 2.1.E in [Ta1]), we have

$$\begin{aligned} \forall v \in H^{\tau,p},\ 1-\delta \leq \tau < 2-\delta &\Longrightarrow A_j^b v \in H^{\tau-1+\delta,p} \\ &\Longrightarrow E^{\#}\partial_j A_j^b v \in H^{\tau+\delta,p}. \end{aligned} \tag{2.12}$$

Next, since $B^j \in L^\infty$, we have

$$u \in H^{1,p} \Longrightarrow E^{\#}B^j\partial_j u \in H^{2,p}. \tag{2.13}$$

It remains to consider the term $E^{\#}Vu$. What we need in order to achieve the regularity result (2.8) after a finite number of iterations and examinations of (2.11) is that

$$E^{\#}V : H^{s,p} \longrightarrow H^{\tau(s),p}, \quad \tau(s) > s, \ s \in [1,2).$$

By interpolation, this will hold whenever (multiplication by) V has the following two properties:

$$V : H^{2,p} \longrightarrow L^p, \tag{2.14}$$

and

$$V : H^{1,p} \longrightarrow H^{-1+\gamma,p}, \quad \gamma = \gamma(p) > 0, \tag{2.15}$$

for p in the range specified in Proposition 2.1. We claim these properties hold whenever $d^{\alpha\beta} \in L^r$, as long as (2.6) and (2.7) hold.

To check (2.14), first assume $p < m/2$. Then, by Sobolev's embedding theorem, $H^{2,p} \subset L^{mp/(m-2p)}$ so, by Hölder's inequality, its image under V is contained in L^s for

$$\frac{1}{s} = \frac{1}{r} + \frac{m-2p}{mp}.$$

If $r > m/2$, the right side is $< 1/p$, so (2.14) holds when $p < m/2$. If $p > m/2$, then $H^{2,p} \subset L^\infty$, and its image under the action of V is included in L^p as long as $r \geq p$. Similarly we can treat $p = m/2$, and the justification of (2.14) is complete.

Now we check (2.15). First assume $p < m$. Then $H^{1,p} \subset L^{mp/(m-p)}$, so its image under V is contained in L^s for

$$\frac{1}{s} = \frac{1}{r} + \frac{m-p}{mp}.$$

If also $p' < m$ (i.e., if $p > m/(m-1)$), so that $H^{1,p'} \subset L^{mp'/(m-p')}$, we will have (2.15) provided

$$\frac{1}{r} + \frac{m-p}{mp} + \frac{m-p'}{mp'} < 1.$$

In fact, the left side is equal to $1/r + 1 - 2/m$, so the desired inequality holds if $r > m/2$. On the other hand, if $p' > m$, so $H^{1,p'} \subset L^\infty$, we have (2.15) provided

$$\frac{1}{r} + \frac{m-p}{mp} < 1,$$

i.e., provided $r > (1 + 1/m - 1/p)^{-1}$. If $p' > m$, then $1 + 1/m - 1/p < 2/m$, so this desired inequality does hold provided $r > m/2$. Similarly one treats $p' = m$, and this disposes of (2.15) when $p < m$.

Now we check (2.15) for $p > m$. In this case $H^{1,p} \subset L^\infty$, so its image under the action of V is contained in L^r. Then we have (2.15) provided

$$\frac{1}{r} + \frac{m-p'}{mp'} < 1,$$

i.e., provided $r > (1 + 1/m - 1/p')^{-1}$. If $p > m$, then $1 + 1/m - 1/p' < 2/m$, so this inequality holds as long as $r > m/2$. This treats (2.15) for $p > m$. Similarly one treats (2.15) for $p = m$.

Thus, under the hypotheses of Proposition 2.1, we have (2.14) and (2.15). In view of the results of (2.11)–(2.13), we deduce that

$$u \in H^{2+\sigma,q} + H^{1+\delta,p} + H^{2,p} + H^{1+\gamma(p),p},$$

under the hypotheses of (2.8). This is an improvement of regularity for u and this argument can be iterated a finite number of times, to produce the conclusion in (2.8). ■

We can now establish an invertibility result.

PROPOSITION 2.2. *Assume that both L and L^* satisfy the hypotheses of Proposition 2.1 and that $r > max\{2, m/2\}$. If L is invertible as a map from $H^{1,2}(\mathcal{M}, \mathcal{E})$ onto $H^{-1,2}(\mathcal{M}, \mathcal{F})$ then, for each $\mu \in [0, 1)$ and $r/(r-1) < p < r$, the map*

$$L : H^{\mu+1,p}(\mathcal{M}, \mathcal{E}) \longrightarrow H^{\mu-1,p}(\mathcal{M}, \mathcal{F}) \tag{2.16}$$

is an isomorphism.

PROOF. By Proposition 2.1, the injectivity of L on $H^{1,2}$ extends to injectivity on $H^{1,p}$, $\forall\, p \geq 2$. For surjectivity, first note that, if $f \in H^{-1,p}(\mathcal{M}, \mathcal{F})$ and $p \in [2, r)$, then there exists $u \in H^{1,2}(\mathcal{M}, \mathcal{E})$ such that $Lu = f$, by hypothesis, and Proposition 2.1 gives $u \in H^{1,p}$. This shows that L is an isomorphism in (2.16), when $\mu = 0$ and $p \in [2, r)$. The result for $p \in (r/(r-1), 2)$ follows by using what we have proved so far for L^* and duality. The invertibility of L in (2.16) for other $\mu \in (0, 1)$ then follows from Proposition 2.1. ■

Proposition 2.2 implies that L has a two-sided inverse

$$E : H^{\mu-1,p}(\mathcal{M}, \mathcal{F}) \longrightarrow H^{\mu+1,p}(\mathcal{M}, \mathcal{E}), \quad 0 \leq \mu < 1, \tag{2.17}$$

given $r > \max\{2, m/2\}$ and $p \in (r/(r-1), r)$. By the Schwartz kernel theorem, E defines an element of $\mathcal{D}'(\mathcal{M} \times \mathcal{M}, \mathcal{E} \otimes \mathcal{F})$ which we also denote E. Assume for a moment that $r > m$ also. Then, since $E(\cdot, y) = E\delta_y$, we have

$$E(\cdot, y) \in H^{1,p}(\mathcal{M}, \mathcal{E} \otimes \mathcal{F}_y), \quad \forall\, p < \frac{m}{m-1}, \text{ uniformly in } y, \tag{2.18}$$

because $r > m/(m-1) > r/(r-1)$ under our hypotheses and $y \mapsto \delta_y$ is continuous.

Since $L_x E(x, y) = 0$ on $\mathcal{M} \setminus \{y\}$, we have from Proposition 2.1:

PROPOSITION 2.3. *Under the hypotheses of Proposition 2.2, if also $r > m$, then*

$$E(\cdot, y) \in H^{s,p}_{\text{loc}}(\mathcal{M} \setminus \{y\}, \mathcal{E} \otimes \mathcal{F}_y), \quad \forall\, s < 2,\ p \in (1, r). \tag{2.19}$$

Hence,

$$E(\cdot, y) \in C^{\gamma}_{\text{loc}}(\mathcal{M} \setminus \{y\}, \mathcal{E} \otimes \mathcal{F}_y), \quad \forall\, \gamma < 2 - \frac{m}{r}.$$

In particular,

$$E \in C^{1+\varepsilon}_{\text{loc}}(\mathcal{M} \times \mathcal{M} \setminus diag, \mathcal{E} \otimes \mathcal{F}), \tag{2.20}$$

for some $\varepsilon = \varepsilon(r, m) > 0$.

We now investigate E on a small neighborhood of the diagonal. Our basic method is similar to that used in the work of M. Mitrea and M. Taylor [MT1]. Given $y_0 \in \mathcal{M}$, let $\mathcal{O}$ be a coordinate neighborhood of y_0, in which L takes the form (2.9). For simplicity, assume $\dim \mathcal{M} = m \geq 3$. Also, in order to facilitate notation we shall occasionally omit the dependence on the vector bundle whenever self evident from the context. Let

$$E_0(D,x) \in \mathrm{\emptyset PC}^1 S^{-2}_{\mathrm{cl}}, \quad E_0(\xi,y) := -\Big(\sum A^{jk}(y)\xi_j\xi_k\Big)^{-1}. \tag{2.21}$$

The Schwartz kernel of $E_0(D,x)$ will be denoted $e_0(x-y,y)$. For later reference, we now pause to record some basic properties of $e_0(z,y)$ which follow more or less directly from (2.21). Concretely, we have

$$e_0(\rho z,y) = |\rho|^{-(m-2)} e_0(z,y) \text{ for any } \rho \in \mathbb{R} \text{ and } z \in \mathbb{R}^m, \tag{2.22}$$

$$(\nabla_2 e_0)(z,y) = \mathcal{O}(|z|^{-(m-2)}) \text{ as } z \to 0 \text{ uniformly in } y, \tag{2.23}$$

and

$$(\nabla_1 e_0)(z,y),\ (\nabla_1\nabla_2 e_0)(z,y) = \mathcal{O}(|z|^{-(m-1)}) \text{ as } z \to 0 \text{ uniformly in } y. \tag{2.24}$$

Returning now to the main line of reasoning, we write

$$\begin{aligned} &LE_0(D,x)u(x) = \\ &(2\pi)^{-m} \iint \big[-A(x,\xi) + B(x,\xi) - V(x)\big] E_0(\xi,y) e^{i(x-y)\cdot\xi} u(y)\,dy\,d\xi, \end{aligned} \tag{2.25}$$

with

$$A(x,\xi) := \xi \cdot A(x)\xi = \sum A^{jk}(x)\xi_j\xi_k, \quad B(x,\xi) := i\sum B^j(x)\xi_j, \tag{2.26}$$

where $A^{jk}(x)$, $B^j(x)$, $V(x)$ are as in the proof of Proposition 2.1. Hence, writing

$$-A(x,\xi)E_0(\xi,y) = I - \xi \cdot \big[A(x) - A(y)\big]\xi\, E_0(\xi,y), \tag{2.27}$$

and setting

$$\begin{aligned} A(x) - A(y) &= \sum H^\ell(x,y)(x_\ell - y_\ell), \\ H^\ell(x,y) &:= \int_0^1 \partial_\ell A(\tau x + (1-\tau)y)\,d\tau, \quad \ell = 1,2,\dots,m, \end{aligned} \tag{2.28}$$

we have, after an integration by parts,

$$LE_0(D,x)u(x) = u(x) + \int R(x,y)u(y)\,dy, \tag{2.29}$$

where

$$\begin{aligned} R(x,y) :=& (2\pi)^{-m} \int [B(x,\xi) - V(x)] E_0(\xi,y) e^{i(x-y)\cdot\xi}\,d\xi \\ &- (2\pi)^{-m} \int \xi \cdot \big[A(x) - A(y)\big]\xi\, E_0(\xi,y) e^{i(x-y)\cdot\xi}\,d\xi \\ =& (2\pi)^{-m} \int [B(x,\xi) - V(x)] E_0(\xi,y) e^{i(x-y)\cdot\xi}\,d\xi \\ &- i(2\pi)^{-m} \int \sum_\ell \frac{\partial}{\partial \xi_\ell} \{(\xi \cdot H^\ell(x,y)\xi) E_0(\xi,y)\} e^{i(x-y)\cdot\xi}\,d\xi. \end{aligned} \tag{2.30}$$

The amplitudes in the last two integrands are sums of terms homogeneous in ξ of degrees -1 and -2. Hence

$$|R(x,y)| \le C|x-y|^{-(m-1)} + C|V(x)|\,|x-y|^{-(m-2)}, \tag{2.31}$$

provided $m \ge 3$. For fixed y, the first term on the right side of (2.31) belongs to L^p_{loc}, for any $p < m/(m-1)$. Also, the second term on the right belongs to L^p_{loc}, as long as

$$\frac{1}{p} > \frac{1}{r} + \frac{m-2}{m}. \tag{2.32}$$

If $r > m$, the right side of this is $< (m-1)/m$, so

$$r > m \Longrightarrow R(\cdot,y) \in L^p_{\mathrm{loc}} \text{ uniformly in } y, \quad \forall\, p < \frac{m}{m-1}. \tag{2.33}$$

We desire to estimate the difference

$$e_1(x,y) = E(x,y)\sqrt{g(y)} - e_0(x-y,y) \tag{2.34}$$

near the diagonal $x = y$. Recall from (1.1) that we are using the convention that, if u is supported on a coordinate patch, then

$$Eu(x) = \int E(x,y)u(y)\sqrt{g(y)}\,dy. \tag{2.35}$$

Recall that we have $L_x e_0(x-y,y) = \delta_y(x) + R(x,y)$ so that

$$L_x e_1(x,y) = -R(x,y). \tag{2.36}$$

We can use Proposition 2.1 to get a preliminary result:

LEMMA 2.4. *Under the hypotheses of Proposition 2.2, if also $r > m$, we have*

$$e_1(\cdot,y) \in H^{s,p}(\mathcal{O}), \quad \forall\; s < 2,\; p < \frac{m}{m-1}, \tag{2.37}$$

$$\nabla_x e_1(\cdot,y) \in L^{p_1}(\mathcal{O}), \quad \forall\; p_1 < \frac{m}{m-2}, \tag{2.38}$$

and

$$e_1(\cdot,y) \in L^{p_2}(\mathcal{O}), \quad \forall\; p_2 < \frac{m}{m-3}, \tag{2.39}$$

uniformly in y.

PROOF. We get (2.37) from (2.36), Proposition 2.1, and the estimate (2.33) on $R(x,y)$. Then (2.38) and (2.39) follow by the Sobolev embedding theorem. ∎

We now look for finer estimates on $e_1(x,y)$. Suppose $|x_0 - y| = 2\rho$. We want to estimate $e_1(x,y)$ on $\{x : |x - x_0| \le \rho/2\}$. Shift coordinates so $x_0 = 0$, and, as in [MT1], introduce the dilation operators

$$v_\rho(x) := v(\rho x), \quad |x| \le 1. \tag{2.40}$$

If we now set $u(x) := e_1(x,y)$ for $|x| \le \rho$ then, from (2.36), u_ρ satisfies the equation

$$\partial_j A^{jk}_\rho \partial_k u_\rho + \rho B^j_\rho \partial_j u_\rho - \rho^2 V_\rho u_\rho = -\rho^2 R_\rho, \tag{2.41}$$

where

$$
\begin{aligned}
A_\rho^{jk}(x) &:= \big(a_{jk}^{\alpha\beta}(\rho x)\big)_{\alpha,\beta}, \quad B_\rho^j(x) := \big(b_j^{\alpha\beta}(\rho x)\big)_{\alpha,\beta}, \\
V_\rho(x) &:= \big(d^{\alpha\beta}(\rho x)\big)_{\alpha,\beta}, \quad R_\rho(x) := R(\rho x, y).
\end{aligned} \tag{2.42}
$$

The equation (2.41) holds on $B_1 = \{x : |x| \leq 1\}$, for ρ in some interval $(0, \rho_0]$, on which the collection A_ρ^{jk} is uniformly Lipschitz, B_ρ^j uniformly bounded, and we have a uniform estimate

$$
\|\rho^2 V_\rho\|_{L^r(B_1)} \leq C, \tag{2.43}
$$

as long as $r \geq m/2$. The family of equations (2.41) is uniformly elliptic. Estimates on the right side of (2.41) follow from (2.30)–(2.31). We have

$$
\|\rho^2 R_\rho\|_{L^r(B_1)} \leq C\rho^{-(m-3)} + C\rho^{-(m-4+m/r)}.
$$

Hence

$$
r > m \Longrightarrow \|\rho^2 R_\rho\|_{L^r(B_1)} \leq C\rho^{-(m-3)}. \tag{2.44}
$$

Now the estimates (2.38)–(2.39) on $e_1(\cdot, y)$ imply for its dilates

$$
\begin{aligned}
\|u_\rho\|_{L^{p_2}(B_1)} &\leq C(p_2)\rho^{-m/p_2}, \quad \forall\, p_2 < \frac{m}{m-3}, \\
\|\nabla u_\rho\|_{L^{p_1}(B_1)} &\leq C(p_1)\rho^{1-m/p_1}, \quad \forall\, p_1 < \frac{m}{m-2}.
\end{aligned} \tag{2.45}
$$

Given the data (2.41), (2.44), and (2.45) on u_ρ, (the proof of) Proposition 2.1 implies

$$
\|u_\rho\|_{H^{s,q}(B_{1/2})} \leq C(s, q, \varepsilon)\rho^{-(m-3+\varepsilon)}, \quad \forall s < 2,\ \varepsilon > 0,\ q < r, \tag{2.46}
$$

provided $r > m$. Hence, for any $\varepsilon > 0$,

$$
\|u_\rho\|_{L^\infty(B_{1/2})} \leq C_\varepsilon \rho^{-(m-3+\varepsilon)}, \quad \|\nabla_x u_\rho\|_{L^\infty(B_{1/2})} \leq C_\varepsilon \rho^{-(m-3+\varepsilon)}. \tag{2.47}
$$

Recalling that $u := e_1(\cdot, y)$, this gives the following.

PROPOSITION 2.5. *If L satisfies the hypotheses of Proposition 2.2 and $r > m$, then for $e_1(x, y) = E(x, y)\sqrt{g(y)} - e_0(x - y, y)$ we have the estimates*

$$
|e_1(x, y)| \leq C_\varepsilon |x - y|^{-(m-3+\varepsilon)}, \quad |\nabla_x e_1(x, y)| \leq C_\varepsilon |x - y|^{-(m-2+\varepsilon)}. \tag{2.48}
$$

Also,

$$
|\nabla_y e_1(x, y)| \leq C_\varepsilon |x - y|^{-(m-2+\varepsilon)}. \tag{2.49}
$$

PROOF. The estimates (2.47) yield (2.48) directly. To get (2.49), let $\tilde{e}_0$ and $\tilde{e}_1$ be constructed in connection with L^*, the formal adjoint of L, in the same way e_0, e_1 have been introduced for L. It follows that

$$
\begin{aligned}
e_1(x, y) &= \sqrt{\frac{g(y)}{g(x)}}\, \tilde{e}_1(y, x)^* + \sqrt{\frac{g(y)}{g(x)}}\, \tilde{e}_0(x - y, x)^* - e_0(x - y, y) \\
&= \sqrt{\frac{g(y)}{g(x)}}\, \tilde{e}_1(y, x)^* + \sqrt{\frac{g(y)}{g(x)}}\, e_0(x - y, x) - e_0(x - y, y),
\end{aligned} \tag{2.50}
$$

where the star superscript denotes the transposed matrix. Since L^* enjoys similar properties to those of L, this and (2.23)–(2.24) easily finish the proof. ■

Note that, if the regularity hypothesis on the coefficients A^{jk} is strengthened to $a_{jk}^{\alpha\beta} \in C^{1+\gamma}$, $\gamma > 0$, then taking δ close enough to 1, we can take $A_j^b \in \mathrm{OP}C^{1+\gamma}S^0_{1,\delta}$ in (2.10). Hence (2.12) holds for $\tau + \delta = 2$, so Proposition 2.1 holds for $\sigma = 0$, and Proposition 2.2 holds for $\mu = 1$. Also (2.19) holds for $s = 2$, and (2.37) holds for $s = 2$. Furthermore, (2.46) holds for $s = 2$, so we can augment the estimates (2.48) by the following:

PROPOSITION 2.6. *In Proposition 2.5, if we add the hypothesis that*

$$a_{jk}^{\alpha\beta} \in C^{1+\gamma}, \quad \textit{for some } \gamma > 0, \tag{2.51}$$

then, for any $\varepsilon > 0$, $q < r$, *we have*

$$\left(\frac{1}{Vol(B_{|x-y|/2}(x))} \int_{B_{|x-y|/2}(x)} |\partial_x^2 e_1(z,y)|^q \, dz \right)^{1/q} \le C_{\varepsilon,q}|x-y|^{-(m-1+\varepsilon)}. \tag{2.52}$$

Let us point out that it is possible to obtain pointwise estimates on $\partial_x^2 e_1(x,y)$ under stronger hypotheses on the coefficients. Specifically, it can be shown that

$$|\partial_x^2 e_1(x,y)| \le C_\varepsilon |x-y|^{-(m-1+\varepsilon)} \tag{2.53}$$

whenever

$$a_{jk}^{\alpha\beta} \in C^{1+\gamma}, \ b_j^{\alpha\beta} \in H^{1,r}, \ d^{\alpha\beta} \in C^\gamma, \quad \text{for some } \gamma \in (0,1), \ r > m. \tag{2.54}$$

This is done using a version of Proposition 2.1 with $\sigma > 0$; as such, this requires $R(x,y)$ to have some smoothness in the variable x and this accounts for the last membership in (2.54). However, we do not develop this point here as we shall only assume weaker conditions than (2.54) in the sequel.

We also want to estimate the first-order x-derivatives of $\nabla_y e_1(x,y)$. We approach this problem as follows. Let Y be a first-order differential operator with smooth coefficients, and consider

$$F := EY = E_0(D,x)Y + E_1 Y. \tag{2.55}$$

The Schwartz kernel of $E_1 Y$ is $f_1(x,y) := Y^t e_1(x,y)$, with Y^t acting in the y-variable. We desire to estimate $f_1(\cdot, y)$ in various norms. Note that, parallel to (2.36), we have

$$L_x f_1(x,y) = -R_Y(x,y), \tag{2.56}$$

where $R_Y(x,y)$ is obtained by applying Y^t to $R(x,y)$, acting in the y-variable. Then,

$$|R_Y(x,y)| \le C|x-y|^{-m} + C|V(x)| \cdot |x-y|^{-(m-1)}. \tag{2.57}$$

which is seen by formally differentiating formula (2.30) for $R(x,y)$. In doing so, care should be taken as we are only assuming $A \in C^{1+\gamma}$. First, when the operator Y_y^t is applied to the second integral in (2.30), there is the contribution $(2\pi)^{-m} \int \xi \cdot (Y^t A)(y)\xi \, E_0(\xi,y) e^{i(x-y)\cdot\xi} \, d\xi$ which is bounded by the first term on the right side of (2.57). When Y_y^t falls on the other factors of the second integral in (2.30), we use (2.28) much the same way as in the second equality in (2.30). Handling the first integral in (2.30) is standard and (2.57) follows in the same spirit to (2.31).

Let us now parallel the analysis in (2.40)–(2.47). Retain the conventions there; say $|x_0-y| = 2\rho$ and shift coordinates so x_0 is the origin. Use the dilation operators (2.40). We want to estimate

$$v_\rho(x) := f_1(\rho x, y) \tag{2.58}$$

on $|x| \le 1/2$. As in (2.41) we have

$$\partial_j A_\rho^{jk} \partial_k v_\rho + \rho B_\rho^j \partial_j v_\rho - \rho^2 V_\rho v_\rho = -\rho^2 \big(R_Y\big)_\rho. \tag{2.59}$$

Parallel to (2.44), we have

$$\|\rho^2 \big(R_Y\big)_\rho\|_{L^r(B_1)} \le C\rho^{-(m-2)}, \tag{2.60}$$

provided $V \in L^r$ and $r > m$. Now, in place of (2.45), we have the following estimate on v_ρ, as a consequence of (2.49), under the hypotheses on Proposition 2.5:

$$\|v_\rho\|_{L^\infty(B_1)} \le C\rho^{-(m-2+\varepsilon)}. \tag{2.61}$$

In order to proceed from here, we need a variant of the regularity result, Proposition 2.1, in which the hypothesis $u \in H^{1,p}_{\text{loc}}(\mathcal{O}, \mathcal{E})$ is replaced by $u \in L^\infty(\mathcal{O}, \mathcal{E})$.

PROPOSITION 2.7. *Let $\mathcal{O} \subset \mathcal{M}$ be open. Assume L is an elliptic operator given by (2.1), with coefficients satisfying*

$$a_{jk}^{\alpha\beta} \in C^{1+\gamma}, \quad b_j^{\alpha\beta} \in H^{1,r}, \quad d^{\alpha\beta} \in L^r. \tag{2.62}$$

Assume

$$\gamma > 0, \quad r > m, \quad r > q > 1, \quad -1 \le \sigma \le 0. \tag{2.63}$$

Then

$$u \in L^\infty_{\text{loc}}(\mathcal{O}, \mathcal{E}),\ Lu \in H^{\sigma,q}_{\text{loc}}(\mathcal{O}, \mathcal{F}) \Longrightarrow u \in H^{2+\sigma,q}_{\text{loc}}(\mathcal{O}, \mathcal{E}) \tag{2.64}$$

and naturally accompanying estimates are valid.

PROOF. Our argument is parallel to the proof of Proposition 2.1 and we make the same notational convention as used there. Note that in the present setting we have $A_j^b \in \text{OP}C^{1+\gamma}S^{1-(1+\gamma)\delta}_{1,\delta} \hookrightarrow \text{OP}C^{1+\gamma}S^0_{1,\delta}$ in the decomposition (2.10), by choosing δ close to 1. Let us examine (2.11) this time. First,

$$u \in L^\infty \Longrightarrow A_j^b u \in \bigcap_{p<\infty} L^p \Longrightarrow E^\# \partial_j A_j^b u \in \bigcap_{p<\infty} H^{1,p}. \tag{2.65}$$

Next, writing $B^j \partial_j u = \partial_j(B^j u) - (\partial_j B^j)u$, we see that

$$u \in L^\infty \Longrightarrow B^j \partial_j u \in L^r + \bigcap_{p<\infty} H^{-1,p} \Longrightarrow E^\# B^j \partial_j u \in H^{2,r} + \bigcap_{p<\infty} H^{1,p}. \tag{2.66}$$

Finally, clearly $u \in L^\infty \Rightarrow E^\# V u \in H^{2,r}$. Thus, the hypotheses of Proposition 2.7 yield

$$u \in H^{2+\sigma,q} + H^{2,r} + \bigcap_{p<\infty} H^{1,p}.$$

This is an improvement of regularity for u. With this at hand, Proposition 2.1 (or by iterating the above scheme) yields the desired conclusion. ■

When Proposition 2.7 is applicable, we can deduce from (2.59)–(2.61) the estimate

$$\|v_\rho\|_{H^{2,q}(B_{1/2})} \leq C(q,\varepsilon)\rho^{-(m-2+\varepsilon)}, \quad \forall\, q < r,\ \varepsilon > 0. \tag{2.67}$$

Hence, for any $\varepsilon > 0$,

$$\|\nabla(v_\rho)\|_{L^\infty(B_{1/2})} \leq C_\varepsilon \rho^{-(m-2+\varepsilon)}. \tag{2.68}$$

Thus we have:

PROPOSITION 2.8. *In Proposition 2.5, strengthen the hypotheses on the coefficients of L to (2.62). Then, in addition to (2.48)–(2.49), $e_1(x,y)$ satisfies*

$$|\nabla_x \nabla_y e_1(x,y)| \leq C_\varepsilon |x-y|^{-(m-1+\varepsilon)}. \tag{2.69}$$

We are now in a position to present the main result of this chapter.

THEOREM 2.9. *Let $\mathcal{E},\mathcal{F} \to \mathcal{M}$ be two (smooth) vector bundles over the (smooth) compact, boundaryless manifold $\mathcal{M}$, of real dimension m. It is assumed that the metric structures on $\mathcal{E}$, $\mathcal{F}$ and $\mathcal{M}$ are Lipschitz continuous.*

Let L be an elliptic, second-order differential operator mapping C^2 sections of $\mathcal{E}$ into measurable sections of $\mathcal{F}$ such that, when written in local coordinates as in (2.1)–(2.2), the coefficients of L satisfy

$$a^{\alpha\beta}_{jk} \in C^1, \quad b^{\alpha\beta}_j \in L^\infty, \quad d^{\alpha\beta} \in L^r, \quad \text{for some } r > m. \tag{2.70}$$

Moreover, suppose that L is invertible as a map from $H^{1,2}(\mathcal{M},\mathcal{E})$ onto $H^{-1,2}(\mathcal{M},\mathcal{F})$ and denote by E the Schwartz kernel of L^{-1}.

Let Ω be an arbitrary Lipschitz domain in $\mathcal{M}$. For a first order differential operator $P \in \mathit{Diff}_1(\mathcal{F},\mathcal{F})$ with continuous coefficients, consider the integral operator with kernel $(Id_x \otimes P_y)E(x,y)$, i.e.,

$$\mathcal{A}f(x) := \int_{\partial\Omega} \langle (Id_x \otimes P_y)E(x,y), f(y)\rangle \, d\sigma(y), \qquad x \in \Omega.$$

Then, if in local coordinates the coefficients of L^* also satisfy (2.70), the following are true. First,

$$\|\mathcal{N}(\mathcal{A}f)\|_{L^p(\partial\Omega)} \leq C_p \|f\|_{L^p(\partial\Omega,\mathcal{F})}, \qquad f \in L^p(\partial\Omega,\mathcal{F}), \quad \forall\, p \in (1,\infty). \tag{2.71}$$

Second, if A is the principal-value boundary integral operator (in the sense of removing geodesic balls) on $\partial\Omega$ with the same kernel, $(Id_x \otimes P_y)E(x,y)$, then

$$A : L^p(\partial\Omega,\mathcal{F}) \longrightarrow L^p(\partial\Omega,\mathcal{E}) \quad \text{is bounded for each } 1 < p < \infty, \tag{2.72}$$

and, third,

$$\mathcal{A}f|_{\partial\Omega_\pm} = \mp\tfrac{1}{2} i\sigma(L)(\cdot,\nu)^{-1}\sigma(P)(\cdot,\nu)^* f + Af \quad \text{a.e. on } \partial\Omega, \tag{2.73}$$

where $*$ denotes here the adjoint.

Going further, assume next that the hypothesis (2.70) for L is strengthened to

$$a^{\alpha\beta}_{jk} \in C^{1+\gamma}, \quad b^{\alpha\beta}_j \in H^{1,r}, \quad d^{\alpha\beta} \in L^r, \quad \text{for some } \gamma > 0,\ r > m. \tag{2.74}$$

Then

$$\mathcal{A}: L^2(\partial\Omega, \mathcal{F}) \longrightarrow H^{1/2,2}(\Omega, \mathcal{E}) \quad \textit{is a bounded operator.} \tag{2.75}$$

Under the hypothesis (2.70) for L and L^*, similar results to (2.71)–(2.72) are valid for the integral operators $\tilde{\mathcal{A}}$ and $\tilde{A}$; these are constructed as before but, this time, in connection with the kernel $(\tilde{P}_x \otimes Id_y)E(x,y)$, where $\tilde{P} \in Diff_1(\mathcal{E}, \mathcal{E})$ has continuous coefficients. In this case, (2.73) becomes

$$\tilde{\mathcal{A}}g|_{\partial\Omega_\pm} = \mp\tfrac{1}{2}i\sigma(\tilde{P})(\cdot,\nu)\sigma(L)(\cdot,\nu)^{-1}g + \tilde{A}g \quad \textit{a.e. on } \partial\Omega, \tag{2.76}$$

for each $g \in L^p(\partial\Omega, \mathcal{E})$, $1 < p < \infty$.

If $\tilde{P}$ has reasonably smooth coefficients, e.g. $H^{1,r}$ for some $r > m$ or C^μ with $\mu > \frac{1}{2}$, and the hypothesis (2.70) for L is strengthened to

$$a_{jk}^{\alpha\beta} \in C^{1+\gamma}, \quad b_j^{\alpha\beta} \in L^\infty, \quad d^{\alpha\beta} \in L^r, \quad \textit{for some } \gamma > 0,\ r > m,$$

then the analogue of (2.75) is also valid in this case. Also, in this case, if Π_L is the volume potential defined by

$$\Pi_L u(x) := \iint_\Omega \langle E(x,y), u(y)\rangle_y \, d\,Vol(y), \qquad x \in \Omega,$$

then

$$\Pi_L : L^p(\Omega, \mathcal{F}) \to H^{2,p}(\Omega, \mathcal{E}) \tag{2.77}$$

is well-defined and bounded for each $1 < p < r$.

PROOF. Using a partition of unity, working in local coordinates (much in the spirit of (2.21)) and finally reassembling the pieces, we can construct

$$Q \in \mathit{OPC}^{1+\gamma}S_{\mathrm{cl}}^{-2}(\mathcal{F}, \mathcal{E}) \tag{2.78}$$

so that $\sigma(Q) = \sigma(L)^{-1}$ and the remainder $R := Q - L^{-1}$ is relatively tame, i.e. its Schwartz kernel $e_1(x,y)$ satisfies (2.48), (2.49), (2.52) and (2.69). It is important to point out here that, granted (2.70) and the fact that the metrics have C^1 coefficients, L^* will automatically satisfy (2.5).

Going further, if $k(Q)$ stands for the Schwartz kernel of Q, the desired conclusions for the contribution of $(\mathrm{Id}_x \otimes P_y)k(Q)$ follows as in Theorem 1.1. Furthermore, as far as (2.71)–(2.73) are concerned, the contribution due to $(\mathrm{Id}_x \otimes P_y)e_1(x,y)$ is, because of the aforementioned estimates, amenable to a simple analysis which we omit. Since the operator with this latter kernel maps $L^2(\partial\Omega, \mathcal{F})$ into $H^{1,2}(\Omega, \mathcal{E})$, by Proposition 2.8, together with Lemma 2.10 which we will establish below, (2.75) also follows.

Essentially, a similar analysis applies to the kernel $(\tilde{P}_x \otimes \mathrm{Id}_y)E(x,y)$ with the notable exception of (2.75). In the case under discussion, this particular estimate follows from the fact that the residual kernel $\partial_x^2 e_1(x,y)$ maps $L^2(\partial\Omega, \mathcal{E})$ boundedly into $H^{s,2}(\Omega, \mathcal{F})$ for any $s < 1$. In turn, this is a consequence of Proposition 2.6 and Lemma 2.11 below. Note that here we also use the well known fact that $H^{1,r} \cdot H^{1/2,2} \hookrightarrow H^{1/2,2}$, etc.

Finally, in order to prove that the operator in (2.77) is bounded for $1 < p < r$, recall the decomposition (2.34) of the kernel $E(x,y)$. The contribution coming from $e_0(x-y,y)$ is handled by performing a decomposition in spherical harmonics and then using the classical Calderón-Zygmund inequality, whereas the contribution coming from the residual part $e_1(x, y)$ is dealt with separately in Lemma 2.12, below (the restriction $p < r$ is inherited from Proposition 2.6).

This completes the proof of Theorem 2.9, modulo the Lemmas 2.10–2.12. ∎

Remarks. (i) In the case when the coefficients of L locally satisfy (2.74) and the metric tensors have components in $H^{2,r}$, then the coefficients of L^* also satisfy (2.74).

(ii) Owing to the parametrix construction earlier in the chapter, in the case when $\mathcal{E} = \mathcal{F}$ and L is also formally self adjoint, $L : H^{1,2}(\mathcal{M}, \mathcal{E}) \to H^{-1,2}(\mathcal{M}, \mathcal{E})$ is an isomorphism if and only if it is one-to-one. (The same conclusion can be reached using Gårding's inequality.)

Here are the lemmas that finish the proof:

LEMMA 2.10. *Let $1 < p < \infty$ and assume $b(x, y)$ satisfies $|b(x, y)| \leq C|x - y|^{-(m-1+\varepsilon)}$, for some $\varepsilon \in (0, \frac{1}{p})$. Then the operator given by*

$$\mathcal{B}f(x) = \int_{\partial\Omega} b(x, y)f(y)\, d\sigma(y) \tag{2.79}$$

maps $L^p(\partial\Omega)$ boundedly into $L^p(\Omega)$.

PROOF. It suffices to prove the analogous result when $\partial\Omega$ is replaced by $\mathbb{R}^{m-1}$, Ω is replaced by $\mathbb{R}^{m-1} \times [0, 1]$ and $b(x, y) = |x - y|^{-(m-1+\varepsilon)}\chi_{\{|x-y|\leq 1\}}$. Now, for each fixed $x \in \mathbb{R}^{m-1} \times [0, 1]$,

$$\int_{\mathbb{R}^{m-1}} |b(x, y)|\, dy \leq C \int_{|x_m|}^{\infty} t^{-(m-1+\varepsilon)}t^{m-2}\, dt = C'|x_m|^{-\varepsilon}. \tag{2.80}$$

Hence, $\int_{\mathbb{R}^{m-1}} |\mathcal{B}f(x', x_m)|^p\, dx' \leq C|x_m|^{-p\varepsilon}\|f\|^p_{L^p(\mathbb{R}^{m-1})}$, and integrating over $x_m \in [0, 1]$ gives the desired estimate, as long as $\varepsilon < 1/p$. ∎

LEMMA 2.11. *Let $1 < p < \infty$ and assume that $b(x, y)$ satisfies, for some $\varepsilon \in (0, \frac{1}{p})$,*

$$(\mathit{Vol}\, B_{\rho/2})^{-1/p}\|b(\cdot, y)\|_{L^p(B_{\rho/2}(x))} \leq C|x - y|^{-(m-1+\varepsilon)}, \tag{2.81}$$

with $\rho := |x - y|$. Then the map (2.79) has the property that

$$\iint_{\Omega} |\mathcal{B}f(x)|^p\, \mathit{dist}\,(x, \partial\Omega)^a\, dx \leq C_{a,p} \int_{\partial\Omega} |f|^p\, d\sigma, \quad \forall\ a > 0. \tag{2.82}$$

PROOF. As in Lemma 2.10, it suffices to prove the analogous result when $\partial\Omega$ is replaced by $\mathbb{R}^{m-1}$ and Ω is replaced by $\mathbb{R}^{m-1} \times [0, 1]$. Also, we can assume that $b(x, y)$ is supported on $|x - y| \leq 1/2$.

For some fixed, arbitrary $h \in (0, 1)$ we will estimate $\iint_{\Gamma_h} |\mathcal{B}f(x)|^p\, dx$, where $\Gamma_h := \mathbb{R}^{m-1} \times [h, 2h]$. To do this, write

$$b(x, y) = \sum_{\nu\geq 0} b_{\nu,h}(x, y), \tag{2.83}$$

with $b_{0,h}(x,y)$ supported on $|x-y| \le h$ and the other terms $b_{\nu,h}(x,y)$ supported on $2^{\nu-1}h \le |x-y| \le 2^\nu h$. The sum ranges over $\{\nu : 2^\nu h \le 1\}$. The corresponding operators will be denoted $\mathcal{B}_{\nu,h}$ so that $\mathcal{B} = \sum_\nu \mathcal{B}_{\nu,h}$. We want to estimate $\mathcal{B}_{\nu,h} f$ on Γ_h. To do this, write

$$\Gamma_h = \bigcup_\alpha Q_{\nu,h,\alpha} \times [h, 2h], \tag{2.84}$$

where $\{Q_{\nu,h,\alpha} : \alpha \in \mathbb{Z}^{m-1}\}$ is a tiling of $\mathbb{R}^{m-1}$ by cubes of edge $2^\nu h$. It follows from Minkowski's inequality, (2.81) and the support conditions that

$$\|\mathcal{B}_{\nu,h} f\|_{L^p(Q_{\nu,h,\alpha}\times[h,2h])} \le C(2^\nu h)^{1-\varepsilon-m/q} \|f\|_{L^1(4Q_{\nu,h,\alpha})}, \tag{2.85}$$

where $1/p + 1/q = 1$ and $4Q_{\nu,k,\alpha}$ is the cube in $\mathbb{R}^{m-1}$ concentric to $Q_{\nu,k,\alpha}$ and having edges four times as large. Thus, $\|f\|_{L^1(4Q_{\nu,h,\alpha})} \le C(2^\nu h)^{(m-1)/q} \|f\|_{L^p(4Q_{\nu,h,\alpha})}$ so that

$$\begin{aligned} \|\mathcal{B}_{\nu,h} f\|_{L^p(\Gamma_h)} &\le C(2^\nu h)^{1/p-\varepsilon} \sum_\alpha \|f\|_{L^p(4Q_{\nu,h,\alpha})} \\ &\le C(2^\nu h)^{1/p-\varepsilon} \|f\|_{L^p(\mathbb{R}^{m-1})}, \end{aligned} \tag{2.86}$$

in view of the way $\{4Q_{\nu,h,\alpha} : \alpha \in \mathbb{Z}^{m-1}\}$ covers $\mathbb{R}^{m-1}$. Hence

$$\begin{aligned} \Big(\iint_{\Gamma_h} |\mathcal{B} f(x)|^p\, dx\Big)^{1/p} &\le \sum_\nu \|\mathcal{B}_{\nu,h} f\|_{L^p(\Gamma_h)} \\ &\le C\Big(\sum_{\{\nu : 2^\nu h \le 1\}} (2^\nu h)^{1/p-\varepsilon}\Big) \|f\|_{L^p(\mathbb{R}^{m-1})} \\ &\le C_\varepsilon \|f\|_{L^p(\mathbb{R}^{m-1})}, \end{aligned} \tag{2.87}$$

with $C_\varepsilon < \infty$ as long as $\varepsilon \in (0, \frac{1}{p})$. Consequently, using this estimate for $h := 2^{-\mu}$ with $\mu = 1, 2, \ldots$, gives

$$\int_0^1 \int_{\mathbb{R}^{m-1}} |\mathcal{B} f(x)|^p x_m^a\, dx \approx \sum_{\mu \ge 1} 2^{-\mu a} \|\mathcal{B} f\|^p_{L^p(\Gamma_{2^{-\mu}})} \le C_a \|f\|^p_{L^p(\mathbb{R}^{m-1})}, \tag{2.88}$$

with $C_a < \infty$ if $a > 0$. ■

LEMMA 2.12. *Assume that $b(x,y)$ is supported on $|x-y| \le 1$ and satisfies*

$$\|b(\cdot, y)\|_{L^p(B_{\rho/2}(x))} \le C|x-y|^{-(m-1+\varepsilon)+m/p}, \qquad \rho := |x-y|, \tag{2.89}$$

for some $p \in (1,\infty)$ and $0 < \varepsilon < 1$. Then the operator

$$\tilde{\mathcal{B}} u(x) := \iint_\Omega b(x,y) u(y)\, dy, \qquad x \in \Omega, \tag{2.90}$$

maps $L^p(\Omega)$ boundedly into itself.

PROOF. Decompose $b(x,y) = \sum_{\nu \leq 0} b_\nu(x,y)$, where $b_\nu(x,y)$ supported on the set $2^{\nu-1} \leq |x-y| \leq 2^\nu$ and, analogously to the proof of Lemma 2.11, write $\tilde{\mathcal{B}} = \sum_{\nu \leq 0} \tilde{\mathcal{B}}_\nu$ for the decomposition induced at the level of operators. Also, assume that $\Omega = \mathbb{R}^m$. Now, if $\{Q_{\nu,\alpha} : \alpha \in \mathbb{Z}^m\}$ is a tiling of $\mathbb{R}^m$ by cubes of edge $2^{\nu-2}$ and $1/p + 1/q = 1$ we have, much as before,

$$
\begin{aligned}
\|\tilde{\mathcal{B}}u\|_{L^p(\mathbb{R}^m)} &\leq \sum_{\nu \leq 0} \|\tilde{\mathcal{B}}_\nu u\|_{L^p(\mathbb{R}^m)} \\
&\leq \sum_{\nu \leq 0} \sum_{\alpha \in \mathbb{Z}^m} \|\tilde{\mathcal{B}}_\nu u\|_{L^p(Q_{\nu,\alpha})} \\
&\leq C \sum_{\nu \leq 0} \sum_{\alpha \in \mathbb{Z}^m} 2^{\nu(1-\varepsilon-m+m/p)} \|u\|_{L^1(100Q_{\nu,\alpha})} \\
&\leq C \sum_{\nu \leq 0} \sum_{\alpha \in \mathbb{Z}^m} 2^{\nu(1-\varepsilon-m+m/p+m/q)} \|u\|_{L^p(100Q_{\nu,\alpha})} \\
&\leq C \Big(\sum_{\nu \leq 0} 2^{\nu(1-\varepsilon)} \Big) \|u\|_{L^p(\mathbb{R}^m)} \\
&\leq C_\varepsilon \|u\|_{L^p(\mathbb{R}^m)}. \qquad (2.91)
\end{aligned}
$$

The proof is finished. ■

We conclude this chapter with another regularity result which will be useful in § 11.

PROPOSITION 2.13. *Retain the same notation and hypotheses as in Theorem 2.9. Also, for an arbitrary, fixed* $1 < p < \infty$, *set*

$$s := min\left\{\tfrac{1}{p}, 1 - \tfrac{1}{p}\right\}. \qquad (2.92)$$

Then, for any $\varepsilon > 0$, *the mapping*

$$\mathcal{A} : L^p(\partial\Omega, \mathcal{F}) \to H^{s-\varepsilon,p}(\Omega, \mathcal{E}) \qquad (2.93)$$

is well defined and bounded.

A similar statement is valid for $\tilde{\mathcal{A}}$ *if the differential operator* $\tilde{P}$ *(see Theorem 2.9) has sufficiently smooth coefficients.*

PROOF. This follows (in the case of $\mathcal{A}$) by interpolating (2.75) with

$$\mathcal{A} : L^p(\partial\Omega, \mathcal{E}) \to L^p(\Omega, \mathcal{E}), \qquad 1 < p < \infty, \qquad (2.94)$$

which is a weaker form of (2.71). Indeed, assume for example that $p \geq 2$ is arbitrary and fixed. Then, given $\varepsilon > 0$, select $p' > \varepsilon^{-1}$ and pick $0 \leq \theta \leq 1$ so that $\dfrac{1}{p} = \dfrac{1-\theta}{2} + \dfrac{\theta}{p'}$. Now $\left[H^{1/2,2}, L^{p'}\right]_\theta = H^{1/p-\mu,p}$ where $\mu = \theta/p' < \varepsilon$ and this completes the proof of this case. The remaining case, $p \leq 2$, is similar and we leave it to the reader. ■

Remark. By further refining the analysis in §§1–2, a sharp form of (2.93) can be proved. Specifically, recalling that $B^{p,q}_s(\Omega)$ is the ordinary Besov scale on Ω, it has been shown in §7 of [MT2] that

$$\mathcal{A} : L^p(\partial\Omega, \mathcal{F}) \to B^{p,p^*}_{1/p}(\Omega, \mathcal{E}), \quad 1 < p < \infty, \tag{2.95}$$

is well defined and bounded, provided the coefficients of L satisfy (2.74). Here and elsewhere, $p^* := \max\{p, 2\}$.

CHAPTER 3

General Second-Order Strongly Elliptic Systems

We continue to assume that the manifold $\mathcal{M}$ and the vector bundles $\mathcal{E}, \mathcal{F} \to \mathcal{M}$ are as in §§ 1–2. Nonetheless, in this chapter we shall restrict attention to the case when $\mathcal{E} = \mathcal{F}$. Consider $L(x, D)$, a second-order differential operator acting on sections of $\mathcal{E}$. Assume $\mathcal{M}$ has a smooth coordinate system, and $\mathcal{E}$ has a local trivialization, with respect to which L has the form

$$Lu = \sum_{j,k} \partial_j A^{jk}(x)\partial_k u + \sum_j B^j(x)\partial_j u - V(x)u, \tag{3.1}$$

where A^{jk}, B^j and V are matrix-valued functions with the following properties:

$$A^{jk} \in C^{1+\gamma}, \quad B^j \in H^{1,r}, \quad V \in L^r, \tag{3.2}$$

for some $\gamma > 0$ and $r > m = \dim \mathcal{M}$. The metric structures on $\mathcal{M}$ and $\mathcal{E}$ are assumed to have C^1 coefficients.

For D a Lipschitz domain in $\mathcal{M}$ (possibly the whole $\mathcal{M}$), we shall say that L satisfies the *non-singularity hypothesis* relative to D provided

$$u \in H_0^{1,2}(D, \mathcal{E}), \ Lu = 0 \text{ in } D \Longrightarrow u = 0 \text{ in } D. \tag{3.3}$$

It is of relevance to remark that, if L is strongly elliptic, then $L - \lambda$, $\lambda \in \mathbb{R}$, satisfies the non-singularity hypothesis (3.3) relative to any subdomain $D \subseteq \mathcal{M}$ provided λ is sufficiently large. This follows from the Gårding inequality, which holds in our setting (despite the fact that V may be unbounded) by (2.15). Also, if L is strongly elliptic and negative semidefinite, then $L - \lambda$ satisfies (3.3) for any $\lambda > 0$.

The main result of this chapter is the theorem below addressing the solvability of the Dirichlet problem for a large class of second order elliptic PDE's in Lipschitz domains. Therefore, from this perspective, this can be considered as a natural extension of the work of D. Jerison and C. Kenig [JK2], which is concerned with the scalar-valued, Euclidean case. However, our approach relies on layer potential methods and is more akin to the works of A. P. Calderón [Ca3], E. Fabes [Fa] (as well as some of their collaborators) whose main results we extend here. Indeed, we produce global representation formulas of the solution by double and single layer potentials. The trend of using such layer potentials "for general elliptic systems" in the nonsmooth context was suggested by A. P. Calderón in [Ca2].

THEOREM 3.1. *Let $\mathcal{E} \to \mathcal{M}$ be as above and consider L a strongly elliptic, (formally) self-adjoint, second-order differential operator acting on sections of $\mathcal{E}$, with coefficients satisfying (3.2) locally. It is also assumed that L satisfies the non-singularity hypothesis (3.3) relative to any Lipschitz subdomain of $\mathcal{M}$.*

Then, for any fixed Lipschitz domain $\Omega \subseteq \mathcal{M}$ there exists $\varepsilon = \varepsilon(\Omega, L) > 0$ such that, for any $f \in L^p(\partial\Omega, \mathcal{E})$, where $2 - \varepsilon < p < 2 + \varepsilon$, the Dirichlet boundary problem

$$\text{(3.4)} \qquad \begin{cases} u \in C^0_{\text{loc}}(\Omega, \mathcal{E}), \\ Lu = 0 \text{ in } \Omega, \\ \mathcal{N}(u) \in L^p(\partial\Omega), \\ u|_{\partial\Omega} = f \text{ on } \partial\Omega, \end{cases}$$

has a unique solution. This solution belongs to $C^\mu_{\text{loc}}(\Omega, \mathcal{E})$ for some $\mu > 1$ and satisfies

$$\text{(3.5)} \qquad \|\mathcal{N}u\|_{L^p(\partial\Omega)} \leq C\|f\|_{L^p(\partial\Omega, \mathcal{E})}$$

for some positive constant C, independent of f.

Also, $u \in H^{1/2,2}(\Omega, \mathcal{E})$ for any $f \in L^2(\partial\Omega, \mathcal{E})$ and

$$\text{(3.6)} \qquad \|u\|_{H^{1/2,2}(\Omega, \mathcal{E})} \leq C\|f\|_{L^2(\partial\Omega, \mathcal{E})}$$

for some $C > 0$ independent of f.

Further, if f actually belongs to $H^{1,p}(\partial\Omega, \mathcal{E})$, then $\mathcal{N}(\nabla u) \in L^p(\partial\Omega)$ and

$$\text{(3.7)} \qquad \|\mathcal{N}(\nabla u)\|_{L^p(\partial\Omega)} \leq C\|f\|_{H^{1,p}(\partial\Omega, \mathcal{E})}$$

for some $C = C(\Omega, L) > 0$. Moreover, if $f \in H^{1,2}(\partial\Omega, \mathcal{E})$ then $u \in H^{3/2,2}(\Omega, \mathcal{E})$ and

$$\text{(3.8)} \qquad \|u\|_{H^{3/2,2}(\Omega, \mathcal{E})} \leq C\|f\|_{H^{1,2}(\partial\Omega, \mathcal{E})}$$

for some $C > 0$ independent of f.

In passing let us point out the obvious corollary that, if L is as in the hypotheses of Theorem 3.1 and if L' is a differential operator such that $L = L'$ in a neighborhood of $\overline{\Omega}$, then the conclusion of this theorem is also valid for L'.

PROOF OF THEOREM 3.1. A construction as in the proof of Proposition 2.1 implies that $L : H^{1,2}(\mathcal{M}, \mathcal{E}) \to H^{-1,2}(\mathcal{M}, \mathcal{E})$ is Fredholm, while the hypotheses of self-adjointness and strong ellipticity allow one to use a deformation argument to imply that L has index zero. Hence, $L : H^{1,2}(\mathcal{M}, \mathcal{E}) \to H^{-1,2}(\mathcal{M}, \mathcal{E})$ is invertible.

Paralleling the arguments in §2, denote by $\Gamma(x, y) \in \mathcal{D}'(\mathcal{M} \times \mathcal{M}, \mathcal{E} \otimes \mathcal{E})$ the Schwartz kernel of $L^{-1} : H^{\mu-1,p}(\mathcal{M}, \mathcal{E}) \to H^{\mu+1,p}(\mathcal{M}, \mathcal{E})$, $0 \leq \mu < 1$, $r/(r-1) < p < r$. Recall from Proposition 2.3 that $\Gamma \in C^{1+\gamma}(\mathcal{M} \times \mathcal{M} \setminus \text{diag}, \mathcal{E} \otimes \mathcal{E})$ for some $\gamma > 0$. Next we introduce the single layer potential

$$\text{(3.9)} \qquad \mathcal{S}f(x) := \int_{\partial\Omega} \langle \Gamma(x, y), f(y) \rangle_y \, d\sigma(y), \quad x \in \mathcal{M} \setminus \partial\Omega,$$

and also set $Sf := \mathcal{S}f|_{\partial\Omega}$.

The idea is to look for a solution to the boundary problem (3.4) expressed in the form

$$\text{(3.10)} \qquad u(x) := \int_{\partial\Omega} \langle (\text{Id} \otimes P)\Gamma(x, y), h(y) \rangle_y \, d\sigma(y) + \int_{\partial\Omega} \langle \Gamma(x, y), \tilde{h}(y) \rangle_y \, d\sigma(y),$$

for $x \in \Omega$, where Id is the identity operator, $P = P(x, D)$ is a suitable first-order differential operator (mapping sections of $\mathcal{E}$ into sections of $\mathcal{E}$), and $h, \tilde{h} \in L^p(\partial\Omega, \mathcal{E})$

are appropriately chosen. Note that for any P, h, $\tilde{h}$, we have $Lu = 0$ in Ω, $\mathcal{N}u \in L^p(\partial\Omega)$ and that, by Theorem 2.9, on the boundary

$$u(x) = -\tfrac{1}{2}i\sigma(L)(x,\nu(x))^{-1}\sigma(P)(x,\nu(x))^*h(x) + Th(x) + S\tilde{h}(x), \tag{3.11}$$

for a.e. $x \in \partial\Omega$, where T is the principal-value singular integral operator on $\partial\Omega$ (in the sense of removing geodesic balls) associated to the kernel $(\mathrm{Id} \otimes P)\Gamma(x,y)$. Thus, in order to prove existence for (3.4) it suffices to show that the operator

$$(h,\tilde{h}) \mapsto -\tfrac{1}{2}i\sigma(L)(\cdot,\nu)^{-1}\sigma(P)(\cdot,\nu)^*h + Th + S\tilde{h} \tag{3.12}$$

maps $L^p(\partial\Omega,\mathcal{E}) \times L^p(\partial\Omega,\mathcal{E})$ onto $L^p(\partial\Omega,\mathcal{E})$ for p close to 2. Now, by Theorem 2.9 and since ontoness is stable on complex interpolation scales (cf. Theorem 2.3 in [KM]), it is enough to prove this only for $p = 2$.

To this end, rewriting the first two terms in (3.12) as

$$\big(-\tfrac{1}{2}i\sigma(L)(\cdot,\nu)^{-1}\sigma(P)(\cdot,\nu)^* + \tfrac{1}{2}(T - T^*)\big) + \tfrac{1}{2}(T + T^*) =: T_1 + T_2$$

we shall show that T_2 is compact on $L^2(\partial\Omega,\mathcal{E})$ and that $P(x,D) \in \mathrm{Diff}_1(\mathcal{E},\mathcal{E})$ can be chosen so that T_1 becomes a strictly accretive operator on $L^2(\partial\Omega,\mathcal{E})$. As far as the latter is concerned, it suffices, granted (2.4), to demand that

$$\sigma(P)(x,\xi) := i\,\xi(w(x))\,\mathrm{Id}_x \quad \text{for } x \in \mathcal{M},\ \xi \in T_x^*\mathcal{M} \setminus 0, \tag{3.13}$$

where $\mathrm{Id}_x : \mathcal{E}_x \to \mathcal{E}_x$ is the identity operator, and $w \in C^\infty(\mathcal{M},T\mathcal{M})$ is a vector field that satisfies

$$\langle w(x), n(x)\rangle_x \geq C > 0 \ \text{ at a.e. } x \in \partial\Omega. \tag{3.14}$$

(Here $n \in L^\infty(\partial\Omega,T\mathcal{M})$ is the outward unit normal to $\partial\Omega$.) A natural candidate for P is the covariant derivative ∇_w, where ∇ is some smooth connection constructed on the vector bundle $\mathcal{E}$. Alternatively, we may invoke the exactness of the short sequence

$$0 \longrightarrow \mathrm{Hom}\,(\mathcal{E},\mathcal{E}) \hookrightarrow \mathrm{Diff}_1(\mathcal{E},\mathcal{E}) \stackrel{\sigma}{\longrightarrow} \mathrm{Hom}(T^*\mathcal{M},\mathrm{Hom}(\mathcal{E},\mathcal{E})) \longrightarrow 0 \tag{3.15}$$

at the tip of the third arrow (this is a version of Theorem 2, pp. 63 in [Pa]) to achieve the same goal.

Next we examine the claim made earlier about the symmetric part of T, i.e. that

$$T_2 = \tfrac{1}{2}(T + T^*) : L^2(\partial\Omega,\mathcal{E}) \longrightarrow L^2(\partial\Omega,\mathcal{E}) \ \text{ is compact.} \tag{3.16}$$

This is going to be a consequence of the weak singularity of its kernel along the diagonal in $\mathcal{M} \times \mathcal{M}$. To see this, we shall work in local coordinates in a small, open neighborhood U of a boundary point $x_0 \in \partial\Omega$. Let $\{\theta_\alpha\}_\alpha$ be an arbitrary local frame in $\mathcal{E}$ over U. As in § 2, in local coordinates, we have the asymptotic expansion

$$\Gamma(x,y) = \frac{1}{\sqrt{g(y)}}\Gamma_0(y,x-y)A(y) + \mathcal{R}(x,y) \tag{3.17}$$

where A is a C^1 matrix-valued function related to the Hermitian structure on $\mathcal{E}$ (in fact, $A := \big((\langle\theta_\alpha,\theta_{\alpha'}\rangle)_{\alpha,\alpha'}\big)^{-1}\big)$ and, for each $\varepsilon > 0$, the matrix-valued residual term $\mathcal{R}$ satisfies

$$|\mathcal{R}(x,y)| + |x-y|\cdot|\nabla_1\mathcal{R}(x,y)| + |x-y|\cdot|\nabla_2\mathcal{R}(x,y)| = \mathcal{O}(|x-y|^{-m+3-\varepsilon}). \tag{3.18}$$

It is of importance to note that the symmetric matrix-valued function $\Gamma_0(y,z)$ satisfies estimates similar to $e_0(z,y)$ in (2.22)–(2.24).

Next, choose $\{\theta_\alpha\}_\alpha$ to be a local orthonormal frame of $\mathcal{E}$ over U in which case A in (3.17) is the identity matrix. Identifying P, constructed above in connection with (3.13), with a matrix of first order differential operators $P \in \mathrm{Diff}_1(U \times \mathbb{R}^{\dim \mathcal{E}}, U \times \mathbb{R}^{\dim \mathcal{E}})$ it follows that the main singularity in the kernel of $T + T^*$ is contained in

$$(3.19) \qquad P(y, D_y)g(y)^{-1/2}\Gamma_0(y, x-y) + (P(x, D_x)g(x)^{-1/2}\Gamma_0(x, y-x))^*,$$

where the star superscript stands here for matrix transposition. Now, since the principal part of P is scalar and since Γ_0 is symmetric, the expression

$$(3.20) \qquad (P(x, D_x)g(x)^{-1/2}\Gamma_0(x, y-x))^* - P(x, D_x)g(x)^{-1/2}\Gamma_0(x, y-x)$$

is actually $\mathcal{O}(|x-y|^{-m+2})$. Therefore, it suffices to examine

$$(3.21) \qquad \begin{aligned} &P(y, D_y)g(y)^{-1/2}\Gamma_0(y, x-y) + P(x, D_x)g(x)^{-1/2}\Gamma_0(x, y-x) \\ &= P(y, D_y)g(y)^{-1/2}\Gamma_0(y, x-y) + P(x, D_x)g(x)^{-1/2}\Gamma_0(x, x-y). \end{aligned}$$

To this end, since $g \in C^1$ and the commutator $[P(\cdot, D), g^{-1/2}]$ is bounded, the above expression can be rewritten as $g^{-1/2}(I + II + III)$ + less singular terms, where

$$(3.22) \qquad \begin{aligned} I &:= P(y, D_y)\Gamma_0(y, x-y) + P(y, D_x)\Gamma_0(y, x-y), \\ II &:= P(x, D_x)\Gamma_0(y, x-y) - P(y, D_x)\Gamma_0(y, x-y), \\ III &:= P(x, D_x)\Gamma_0(x, x-y) - P(x, D_x)\Gamma_0(y, x-y); \end{aligned}$$

(recall that $\Gamma_0(x,z)$ is even in z). On account of the smoothness of the coefficients of P and the estimates (2.23)–(2.24), we obtain that each of the quantities above and, hence, the entire expression (3.19), is $\mathcal{O}(|x-y|^{-m+2-\varepsilon})$ for any $\varepsilon > 0$. Consequently, the proof of (3.16) is finished by invoking (3.18).

At this point we may conclude that the operator $-\frac{1}{2}i\sigma(L)(\cdot,\nu)^{-1}\sigma(P)(\cdot,\nu)^* + T$ is Fredholm with index zero. In particular, it has a finite codimensional range. Next, we proceed to showing that

$$(3.23) \qquad S : L^2(\partial\Omega, \mathcal{E}) \to L^2(\partial\Omega, \mathcal{E}) \text{ has dense range.}$$

Since L is formally self-adjoint we have that $\Gamma(x,y)$ is symmetric, i.e., $\Gamma(y,x) = \Gamma(x,y)^*$, where the star denotes the transposition isomorphism $\mathcal{E}_x \otimes \mathcal{E}_y \simeq \mathcal{E}_y \otimes \mathcal{E}_x$. Consequently, S is a self-adjoint mapping of $L^2(\partial\Omega, \mathcal{E})$ and, hence, (3.23) is equivalent to

$$(3.24) \qquad S : L^2(\partial\Omega, \mathcal{E}) \to L^2(\partial\Omega, \mathcal{E}) \text{ is injective.}$$

To prove this, let $h \in L^2(\partial\Omega, \mathcal{E})$ be arbitrary with $Sh = 0$ on $\partial\Omega$, and set $u := \mathcal{S}h$ in $\mathcal{M} \setminus \partial\Omega$. Because $u \in H^{1,2}(\Omega_\pm, \mathcal{E})$ satisfies $Lu = 0$ in $\mathcal{M} \setminus \partial\Omega$ and $u|_{\partial\Omega_\pm} = 0$, the non-singularity hypothesis implies that $u = 0$ in $\mathcal{M} \setminus \partial\Omega$. In particular, if $P(x,D) \in \mathrm{Diff}_1(\mathcal{E}, \mathcal{E})$ is as before then

$$(3.25) \qquad P(x, D)u = 0 \text{ in } \mathcal{M} \setminus \partial\Omega.$$

Going nontangentially to the boundary in (3.25) from both sides and subtracting the two traces yields

$$(3.26) \qquad 0 = P(x,D)u|_{\partial\Omega_-} - P(x,D)u|_{\partial\Omega_+} = i\sigma(P)(x,\nu(x))\sigma(L)(x,\nu(x))^{-1}h.$$

Hence, $h = 0$ by (2.4), (3.13) and (3.14), proving (3.24). At this stage we can conclude that

$$\text{Range}\,(S) + \text{Range}\,(-\tfrac{1}{2}i\sigma(L)(\cdot,\nu)^{-1}\sigma(P)(\cdot,\nu)^* + T) = L^2(\partial\Omega,\mathcal{E}),$$

i.e., that the operator (3.12) is onto for p close to 2. This finishes the proof of the existence part for (3.4).

Next we shall prove that if $|2-p| \le \varepsilon$ (with ε perhaps smaller than before) then

$$S : L^p(\partial\Omega,\mathcal{E}) \to H^{1,p}(\partial\Omega,\mathcal{E}) \text{ is an isomorphism.} \tag{3.27}$$

Once again, it suffices to treat the case $p = 2$ only. An incisive step in this direction is establishing the *a priori* estimate

$$\int_{\partial\Omega} |P(x,D)u|^2\,d\sigma \le C\|f\|^2_{H^{1,2}(\partial\Omega,\mathcal{E})} + C\|u\|^2_{H^{1,2}(\Omega,\mathcal{E})}, \tag{3.28}$$

where $C = C(\partial\Omega) > 0$, for any solution of $Lu = 0$ in Ω, $u\big|_{\partial\Omega} = f$, with appropriate control at the boundary. To this end, observe that it is enough to prove a local version of (3.28) which we now proceed to describe. Let Ω be a Lipschitz domain in $\mathbb{R}^m$ whose surface measure, inherited from the Euclidean metric, is denoted by $d\sigma$ (note that if ρ is the Radon-Nikodym derivative of this and the surface measure induced by the Riemannian metric then $\rho,\rho^{-1} \in L^\infty(\partial\Omega)$). Next, let U, W be two small, bounded, open neighborhoods of a boundary point $x_0 \in \partial\Omega$ so that $\bar{U} \subseteq W$. Consider $L = (L_{\alpha\beta})_{\alpha,\beta}$ a matrix of second-order differential operators as in (2.2), with principal symbol $\sigma(L)(x,\xi) = \left(-\sum_{j,k} a^{\alpha\beta}_{jk}(x)\xi_j\xi_k\right)_{\alpha,\beta}$. The strong ellipticity assumption on L translates into this matrix being negative definite, i.e.

$$\sum_{j,k}\sum_{\alpha,\beta} a^{\alpha\beta}_{jk}(x)\xi_j\xi_k\eta^\alpha\eta^\beta \ge C|\xi|^2|\eta|^2 \tag{3.29}$$

for any $\xi \in \mathbb{R}^m$, $\eta \in \mathbb{R}^{\dim\mathcal{E}}$, uniformly in $x \in W$, whereas the self-adjointness of L gives that

$$\left(\sum_{j,k} a^{\alpha\beta}_{jk}(x)\xi_j\xi_k\right)_{\alpha,\beta} \quad \text{is symmetric} \tag{3.30}$$

for any $x \in W$. Then, so we claim, there exists a constant $C = C(\partial\Omega, L, U, W) > 0$ such that

$$\begin{aligned}\int_{U\cap\partial\Omega} |\nabla u|^2\,d\sigma \le{} & C\int_{W\cap\partial\Omega} |\nabla_{\tan} u|^2\,d\sigma + C\int_{W\cap\partial\Omega}|u|^2\,d\sigma \\ & + C\iint_{W\cap\Omega}\left(|\nabla u|^2 + |u|^2\right)d\mathrm{Vol}\end{aligned} \tag{3.31}$$

for any $u = (u^\alpha)_\alpha$ solution of $Lu = 0$ in $\Omega\cap W$ with $\mathcal{N}(\nabla u) \in L^2(\partial\Omega\cap W)$. Here $|\nabla u| := \left(\sum_{\alpha,j}|\partial_j u^\alpha|^2\right)^{1/2}$ and $\nabla_{\tan}$ is the tangential gradient on $\partial\Omega$ (acting component-wise).

To see this, consider $h = (h_j)_j$ a smooth vector field in $\mathbb{R}^m$ with support in W and such that, if n denotes the (Euclidean) outgoing unit normal to $\partial\Omega$,

$$\langle h(x), n(x)\rangle \ge C > 0 \text{ a.e. on } U\cap\partial\Omega. \tag{3.32}$$

We shall make use of a variable-coefficient version of a Rellich type identity (due to J. Nečas [Ne] and L. Payne and H. Weinberger [PW]; cf. also C. Kenig [Ke1]) which, in our case, reads

$$
\begin{aligned}
&\sum_{\alpha,\beta,j,k,l} \frac{\partial}{\partial x_l}\left[\left(h_l(x)a_{jk}^{\alpha\beta}(x) - h_j(x)a_{lk}^{\alpha\beta}(x) - h_k(x)a_{jl}^{\alpha\beta}(x)\right)\frac{\partial u^\alpha}{\partial x_j}(x)\frac{\partial u^\beta}{\partial x_k}(x)\right] \\
&\qquad = -2\sum_{l,\alpha} h_l(x)\frac{\partial u^\alpha}{\partial x_l}(x)\left(\sum_{\beta,j,k} a_{jk}^{\alpha\beta}(x)\frac{\partial^2 u^\beta}{\partial x_j \partial x_k}(x)\right) \\
&\qquad\quad + \mathcal{O}\big(|\nabla u|^2[|\nabla h| + |h|]\big).
\end{aligned} \tag{3.33}
$$

This can be verified by a straightforward calculation which utilizes the symmetry condition (3.30). Going further, note that since $Lu = 0$, we may replace $\sum_{\beta,j,k} a_{jk}^{\alpha\beta}(x)\frac{\partial^2 u^\beta}{\partial x_j\partial x_k}(x)$ by the lower-order expression

$$
\begin{aligned}
&-\sum_{\beta,j,k}\frac{\partial a_{jk}^{\alpha\beta}}{\partial x_j}(x)\frac{\partial u^\beta}{\partial x_k}(x) - \sum_{\beta,j} b_j^{\alpha\beta}(x)\frac{\partial u^\beta}{\partial x_j}(x) - \sum_\beta d^{\alpha\beta}(x)u^\beta(x) \\
&= \mathcal{O}(|\nabla u| + |u|) + \mathcal{O}\big(|V||u|\big),
\end{aligned} \tag{3.34}
$$

where $V := (d^{\alpha\beta})_{\alpha,\beta}$. With this in mind, the identity (3.33), the divergence theorem and simple algebraic manipulations based on the symmetry condition (3.30) readily imply that

$$
\begin{aligned}
&\int_{\partial\Omega}\langle h(x), n(x)\rangle\left(\sum_{\beta,j,k} a_{jk}^{\alpha\beta}(x)\frac{\partial u^\alpha}{\partial x_j}(x)\frac{\partial u^\beta}{\partial x_k}(x)\right)d\sigma(x) \\
&=2\int_{\partial\Omega}\sum_{\alpha,\beta,l,j,k}\left(n_l(x)h_l(x)a_{jk}^{\alpha\beta}(x) - n_l(x)h_j(x)a_{lk}^{\alpha\beta}(x)\right)\frac{\partial u^\alpha}{\partial x_j}(x)\frac{\partial u^\beta}{\partial x_k}(x)\,d\sigma(x) \\
&\quad + \iint_{\Omega\cap W}\mathcal{O}(|\nabla u|^2 + |u|^2)\,d\mathrm{Vol} + \iint_{\Omega\cap W}\mathcal{O}\big(|\nabla u||u||V|\big)\,d\mathrm{Vol}.
\end{aligned} \tag{3.35}
$$

Note that $u \in H^{1,2} \subset L^q$ provided $1/q \geq 1/2 - 1/m$. In particular, since $V \in L^r$ with $r > m$, matters can be arranged so that $1/2 + 1/q + 1/r = 1$. By the generalized Hölder's inequality, it follows that the last integral in (3.35) is majorized by a multiple of $\|u\|^2_{H^{1,2}(\Omega\cap W)}$ and, hence, it can be absorbed in the previous integral in (3.35).

Next, a purely algebraic argument based on (3.29) gives the pointwise estimate

$$
|\nabla u(x)|^2 \leq C\left(\sum_{\beta,j,k} a_{jk}^{\alpha\beta}(x)\frac{\partial u^\alpha}{\partial x_j}(x)\frac{\partial u^\beta}{\partial x_k}(x)\right) + C|\nabla_{\tan} u|^2(x) \tag{3.36}
$$

almost everywhere on $\partial\Omega \cap V$. Indeed, this is immediate by writing $\partial_j u^\alpha = n_j\frac{\partial u^\alpha}{\partial n} + \mathcal{O}(|\nabla_{\tan} u|)$ plus a similar decomposition for $\partial_k u^\beta$ and then using the positive definiteness of the matrix $a_{jk}^{\alpha\beta}(x)n_j n_k$ uniformly in x. If we now observe that for each fixed k, α, β,

$$\sum_{l,j} \Big(n_l(x) h_l(x) a_{jk}^{\alpha\beta}(x) - n_l(x) h_j(x) a_{lk}^{\alpha\beta}(x) \Big) \frac{\partial}{\partial x_j}$$

is a tangential derivative on $\partial\Omega$, then (3.31) follows from (3.35), (3.36) and (3.32). This completes the proof of (3.28).

Proceeding further, for an arbitrary $f \in L^2(\partial\Omega, \mathcal{E})$, set $u := \mathcal{S}f$ in $\mathcal{M} \setminus \partial\Omega$. Then, by the accretivity of $-i\,\sigma(P)(\cdot,\nu)\sigma(L)(\cdot,\nu)^{-1}$, the jump relations for $P(x,D)\mathcal{S}f$ and (3.28) written both for Ω_- and Ω_+ we have, for some $C = C(\partial\Omega) > 0$,

$$\begin{aligned} \|f\|_{L^2(\partial\Omega,\mathcal{E})} &\le C\| - i\sigma(P)(\cdot,\nu)\sigma(L)(\cdot,\nu)^{-1} f\|_{L^2(\partial\Omega,\mathcal{E})} \\ &\le C\|P(x,D)u|_{\partial\Omega_+}\|_{L^2(\partial\Omega,\mathcal{E})} + C\|P(x,D)u|_{\partial\Omega_-}\|_{L^2(\partial\Omega,\mathcal{E})} \\ &\le C\|Sf\|_{H^{1,2}(\partial\Omega,\mathcal{E})} + C\|\mathcal{S}f\|_{H^{1,2}(\Omega_+,\mathcal{E})} + C\|\mathcal{S}f\|_{H^{1,2}(\Omega_-,\mathcal{E})}. \end{aligned} \tag{3.37}$$

In particular, since $\mathcal{S} : L^2(\partial\Omega, \mathcal{E}) \to H^{1,2}(\Omega_\pm, \mathcal{E})$ are compact (e.g., by Theorem 2.9), this implies that S is bounded from below modulo compact operators and, hence,

$$S : L^2(\partial\Omega, \mathcal{E}) \to H^{1,2}(\partial\Omega, \mathcal{E}) \text{ has a closed range.} \tag{3.38}$$

In order to show that this operator has also a dense range, we shall assume for a moment that $\partial\Omega$ is a *smooth* submanifold of codimension one in $\mathcal{M}$, and recall that the leading part in L has C^μ coefficients for some $\mu > 1$ and, also, recall from (2.78) the pseudodifferential "approximation" Q of L^{-1}. Further, denote by $\tilde{S}$ the single layer potential operator associated with the Schwartz kernel of Q. Using (2.49), (2.69), and a simple interpolation argument involving operators with integral kernel $k_\zeta(x,y) := \text{dist}(x,y)^\zeta \nabla_y e_1(x,y)$, for Re $\zeta > -1$, we deduce that

$$S - \tilde{S} : H^{-1,2}(\partial\Omega, \mathcal{E}) \to H^{1-\varepsilon,2}(\partial\Omega, \mathcal{E}), \quad \forall\, \varepsilon > 0. \tag{3.39}$$

Proceeding similarly or, alternatively, using (3.39), duality and the estimates in §2, we can also prove that

$$S - \tilde{S} : H^{-1+\varepsilon,2}(\partial\Omega, \mathcal{E}) \to H^{1,2}(\partial\Omega, \mathcal{E}), \quad \forall\, \varepsilon > 0. \tag{3.40}$$

Next, we claim that the mapping $f \mapsto \tilde{S}f = Q(\sigma f)|_{\partial\Omega}$, regarded as a pseudodifferential operator on $\partial\Omega$, i.e.

$$\tilde{S} \in \mathcal{O}\mathrm{P}C^\mu S^{-1}_{1,0}(\mathcal{E}|_{\partial\Omega}, \mathcal{E}|_{\partial\Omega}), \tag{3.41}$$

is strongly elliptic. Here, for each smooth section f on $\partial\Omega$, $\sigma f \in \mathcal{D}'(\mathcal{M}, \mathcal{E})$ is the distribution on $\mathcal{M}$ given by $C^\infty(\mathcal{M}, \mathcal{E}) \ni \phi \mapsto \int_{\partial\Omega} \langle f(x), \phi(x) \rangle_x \, d\sigma(x)$. The claim seems to be well-known, at least for various special cases, and here we outline a brief argument. First, for $x \in \partial\Omega$ and $\xi \in T_x^*\partial\Omega$, $\xi \neq 0$, we have that

$$\begin{aligned} \sigma(\tilde{S})(x,\xi) &= C_m \int_{-\infty}^{+\infty} \sigma(Q)(x, \xi + t\nu(x))\, dt \\ &= C_m \int_{-\infty}^{+\infty} \sigma(L)(x, \xi + t\nu(x))^{-1}\, dt \end{aligned} \tag{3.42}$$

(here we regard $\xi \in T_x^*\partial\Omega$ as an element in $T^*\mathcal{M}$ by setting $\langle \xi, \nu \rangle = 0$). The first equality is seen by locally flattening $\partial\Omega$ (much in the spirit of (11.11)–(11.12) in

Chapter 7 in [Ta2]), whereas the second one follows from the construction of Q. Note that $\sigma(\tilde{S})(x,\xi)$ is positive homogeneous of degree -1 in ξ and that the integrals above are absolutely convergent as $|\sigma(L)(x,\xi+t\nu(x))^{-1}|_{\mathrm{Hom}(\mathcal{E}_x,\mathcal{E}_x)} \leq C(|\xi|^2+t^2)^{-1}$. Then, for any $\eta \in \mathcal{E}_x$, and $\xi \in T^*_x\partial\Omega$,

$$\begin{aligned} \langle -\sigma(\tilde{S})(x,\xi)\eta,\eta\rangle_x &= C\int_{-\infty}^{+\infty} \langle -\sigma(L)(x,\xi+t\nu(x))^{-1}\eta,\eta\rangle_x\,dt \\ &\geq C|\eta|^2\int_{-\infty}^{+\infty}(|\xi|^2+t^2)^{-1}\,dt \\ &\geq C|\eta|^2|\xi|^{-1}, \end{aligned} \tag{3.43}$$

for some $C > 0$. This yields the strong ellipticity of $\tilde{S}$.

Next, we continue to assume $\partial\Omega \in C^\infty$ and aim at proving that

$$S : H^{-1,2}(\partial\Omega,\mathcal{E}) \to L^2(\partial\Omega,\mathcal{E}) \text{ is injective.} \tag{3.44}$$

First we establish some regularity of any element of the kernel. Parallel to (2.10), we decompose $\tilde{S} \in \mathrm{OP}C^\mu S^{-1}_{1,0}$ into two pieces $\tilde{S} = \tilde{S}^\# + \tilde{S}^b$ where $\tilde{S}^\#$, the "sharp" piece, belonging to $\mathrm{OP}S^{-1}_{1,\delta}(\mathcal{E}|_{\partial\Omega},\mathcal{E}|_{\partial\Omega})$, has a smooth symbol, and the "bad" piece $\tilde{S}^b$, belonging to $\mathrm{OP}C^\mu S^{-1-\delta\mu}_{1,\delta}(\mathcal{E}|_{\partial\Omega},\mathcal{E}|_{\partial\Omega})$, has a rough but lower-order symbol. Here, $S^\ell_{\rho,\delta}$ is the usual Hörmander class of symbols, and we choose $\delta \in (0,1)$ so that $-1+\delta\mu \geq 0$.

Let $E \in \mathrm{OP}S^1_{1,\delta}(\mathcal{E}|_{\partial\Omega},\mathcal{E}|_{\partial\Omega})$ be a parametrix for $\tilde{S}^\#$ so that $E\tilde{S}^\# = \mathrm{Id} + R$ with R a smoothing operator. Then, if $f \in H^{-1,2}(\partial\Omega,\mathcal{E})$ is such that $Sf = 0$, it follows that

$$f = -E\tilde{S}^b f - Rf + E(\tilde{S} - S)f. \tag{3.45}$$

Invoking results of G. Bourdaud [Bou], we have that $\tilde{S}^b$ maps $H^{-1,2}(\partial\Omega,\mathcal{E})$ boundedly into $H^{\delta\mu,2}(\partial\Omega,\mathcal{E})$ so that $E\tilde{S}^b f \in H^{-1+\delta\mu,2}(\partial\Omega,\mathcal{E})$. Based on this, the dual of (3.40) and our choice of δ , we obtain with the help of (3.45) that $f \in H^{-\varepsilon,2}(\partial\Omega,\mathcal{E})$ for any $\varepsilon > 0$. Substituting this back in (3.45) and invoking (3.39), we finally arrive at $f \in L^2(\partial\Omega,\mathcal{E})$. Thus, at this point, (3.24) implies $f = 0$ and this finishes the proof of (3.44).

Dualizing (3.44) gives that $S : L^2(\partial\Omega,\mathcal{E}) \to H^{1,2}(\partial\Omega,\mathcal{E})$ has dense range. Summarizing, from this and (3.38) we may conclude that

$$\partial\Omega \in C^\infty \Longrightarrow S : L^2(\partial\Omega,\mathcal{E}) \to H^{1,2}(\partial\Omega,\mathcal{E}) \text{ is invertible.} \tag{3.46}$$

We now observe that the extra smoothness assumption on $\partial\Omega$ may be eliminated, i.e. we may return to the usual assumption that $\partial\Omega$ is only Lipschitz continuous. This may be done by a semi-standard limiting argument which utilizes (3.24), (3.46) and the fact that the constant C in (3.37) depends only on the Lipschitz character of Ω. Ultimately, this yields

$$\{F|_{\partial\Omega};\ F \in C^\infty(\mathcal{M},\mathcal{E})\} \subseteq \mathrm{Range}\,(S) \tag{3.47}$$

for any Lipschitz domain Ω; see, e.g., [Ve1] or [MT1] for details in fairly similar circumstances.

Therefore, (3.27) follows from (3.24), (3.38) and (3.47). In particular, if $f \in H^{1,p}(\partial\Omega)$, p close to 2, then

$$u(x) := \mathcal{S}(S^{-1}f)(x) \tag{3.48}$$

solves (3.4) and this takes care of the regularity statement (3.7) in Theorem 3.1.

Turning to the uniqueness part, we introduce the Green function

$$G(x,y) := \Gamma(x,y) - \mathrm{Id}\otimes\mathcal{S}\left[\mathrm{Id}\otimes S^{-1}\left(\Gamma(x,\cdot)|_{\partial\Omega}\right)\right](y), \quad (x,y)\in\Omega\times\Omega\setminus\mathrm{diag}. \tag{3.49}$$

Note that $\Gamma(x,\cdot)|_{\partial\Omega}$ belongs to $\mathcal{E}_x\otimes H^{1,p}(\partial\Omega,\mathcal{E})$ for any $1<p<\infty$ and $x\in\Omega$, whereas $\mathrm{Id}\otimes S^{-1}:\mathcal{E}_x\otimes H^{1,p}(\partial\Omega,\mathcal{E})\to\mathcal{E}_x\otimes L^p(\partial\Omega,\mathcal{E})$ is well defined for p close to 2. Thus, $G(x,y)\in\mathcal{E}_x\otimes\mathcal{E}_y$ is well defined.

Now, consider a sequence of Lipschitz subdomains Ω_j of Ω so that $\Omega_j\nearrow\Omega$ as $j\to\infty$ (cf. Appendix A in [MT1]) and such that their Lipschitz character is controlled uniformly in j. Set G_j for the Green function corresponding to Ω_j. By construction, $G_j(x,\cdot)|_{\partial\Omega_j}=0$ and we claim that also

$$\sup_j \|\mathcal{N}_j(\nabla_2 G_j(x,\cdot))\|_{L^p(\partial\Omega_j)} \le C < +\infty \tag{3.50}$$

for p close to 2. This follows from the fact that the operator norm of S_j^{-1} : $H^{1,p}(\partial\Omega_j,\mathcal{E})\to L^p(\partial\Omega_j,\mathcal{E})$ is uniformly bounded in j. In turn, this is seen from (3.37), stability results on complex interpolation scales (cf. the remark preceding Theorem 2.7 in [KM]), and reasoning by contradiction.

Integrations by parts against these Green functions give that for any solution u of the homogeneous version of (3.4) and any $x\in\Omega$,

$$\begin{aligned} u(x) &= \iint_{\Omega_j} \langle (LG_j(x,\cdot))(y), u(y)\rangle_y \, d\mathrm{Vol}(y) \\ &= \int_{\partial\Omega_j} \mathcal{O}(|u|\cdot|\nabla_2 G_j(x,\cdot)|)\, d\sigma_j \\ &\le C\|u\|_{L^p(\partial\Omega_j,\mathcal{E})}, \end{aligned} \tag{3.51}$$

where the last inequality utilizes (3.48). Because $\|u\|_{L^p(\partial\Omega_j,\mathcal{E})}\to 0$ by Lebesgue's dominated convergence theorem, we get $u(x)=0$ and since $x\in\Omega$ was arbitrary the uniqueness statement follows.

Finally, (3.5) and (3.6) are consequences of (3.10), (3.12), Theorem 2.3 in [KM] and Theorem 2.9, whereas (3.7) and (3.8) follow from (3.48) and Theorem 2.9. This completes the proof of Theorem 3.1. ∎

We complement the statement of Theorem 3.1 with several important observations.

REMARK I. The existence of the (non-tangential) boundary values in the formulation of the boundary problem (3.4) can be made part of the conclusion of the Theorem 3.1. Indeed, so we claim, any $u\in C^0(\Omega,\mathcal{E})$ with $Lu=0$ in Ω and $\mathcal{N}u\in L^p(\partial\Omega)$, $2-\varepsilon<p<2+\varepsilon$, has a well-defined trace $u|_{\partial\Omega}$ (in fact, $\mathcal{N}u\in L^p(\partial\Omega)$ can be replaced by uniform L^p-integrability on a sequence of "parallel" boundaries). This is an immediate corollary of Theorem 2.9 and the fact that, under these hypotheses, u admits the integral representation formula (3.10) in Ω. In turn, the latter follows from a similar integral representation formula in a sequence of (suitably chosen) exhausting domains $\Omega_j \subset\subset \Omega$, i.e.

$$
\begin{aligned}
u(x) := & \int_{\partial\Omega_j} \langle (\mathrm{Id} \otimes P)\Gamma(x,y), h_j(y)\rangle_y \, d\sigma_j(y) \\
& + \int_{\partial\Omega_j} \langle \Gamma(x,y), \tilde{h}_j(y)\rangle_y \, d\sigma_j(y), \quad x \in \Omega_j,
\end{aligned} \tag{3.52}
$$

valid for certain $h_j, \tilde{h}_j \in L^p(\partial\Omega_j, \mathcal{E})$ with

$$
\|h_j\|_{L^p(\partial\Omega_j,\mathcal{E})} + \|\tilde{h}_j\|_{L^p(\partial\Omega_j,\mathcal{E})} \leq C\|\mathcal{N}u\|_{L^p(\partial\Omega)}, \tag{3.53}
$$

with C independent of j, plus an elementary weak$-*$ argument. The sequence of approximating Lipschitz domains Ω_j are constructed by selecting a C^1-smooth vector field w in $\mathcal{M}$ which is transversal to $\partial\Omega$ and then discarding from Ω all points lying on orbits $\gamma(x,t)$ of w emerging (at the initial moment) inward from boundary points $x \in \partial\Omega$, where the time parameter t is restricted to an interval of the form $[0, 1/j]$ (for j large enough). Clearly, (3.52) is a consequence of Theorem 3.1 since $u \in C^0(\overline{\Omega}_j, \mathcal{E})$. Further, (3.53) follows from the fact that $T_j \to T$, $S_j \to S$ in the operator norm (cf. the work of A. P. Calderón [Ca3] for a similar claim; here the subindex j labels objects constructed in connection with $\partial\Omega_j$–recall that T has been introduced in connection with (3.11)–and we identify $\partial\Omega_j$ with $\partial\Omega$ by matching points on the same orbits), together with the following functional analytic result: If $A : X \to Y$ is a bounded operator from a Banach space X onto another Banach space Y and if $(A_j)_j$ is a sequence of bounded operators convergent to A in the operator norm, then there exists $C > 0$ and j_0 such that

$$
\forall j \geq j_0, \ \forall y \in Y \Longrightarrow \exists x \in X \text{ so that } A_j x = y, \ \|x\| \leq C\|y\|. \tag{3.54}
$$

This is a consequence of the open mapping theorem. For large j, we always have "good" approximate solutions to $A_j x = y$ and this, by an iteration argument, gives a nearby actual solution.

REMARK II. It is useful to explicitly point out a consequence of the integral representation formula (3.10) and the estimates in §§ 1 − 2. Specifically, we have

$$
\iint_\Omega |\nabla u(x)|^2 \mathrm{dist}\,(x, \partial\Omega)\, d\mathrm{Vol} \leq C\|u\|^2_{L^2(\partial\Omega,\mathcal{E})}, \tag{3.55}
$$

uniformly for $u \in C^0(\Omega, \mathcal{E})$ satisfying $Lu = 0$ in Ω and $\mathcal{N}(u) \in L^2(\partial\Omega)$. Introducing the so-called area-function

$$
\mathcal{A}(u)(x) := \left(\iint_{\gamma(x)} |\nabla u|^2 \, \mathrm{dist}\,(\cdot, \partial\Omega)^{2-m} \, d\mathrm{Vol} \right)^{1/2}, \qquad x \in \partial\Omega,
$$

where $\gamma(x) \subset \Omega$ stands for the nontangential approach region at x, the estimate (3.55) readily implies that

$$
\|\mathcal{A}(u)\|_{L^2(\partial\Omega)} \leq C\|\mathcal{N}(u)\|_{L^2(\partial\Omega)}.
$$

Furthermore, this and fairly standard arguments (see §4 in [DKPV]) yield the L^p-version of this latter estimate for any $0 < p < \infty$. That is,

$$
\|\mathcal{A}(u)\|_{L^p(\partial\Omega)} \leq C\|\mathcal{N}(u)\|_{L^p(\partial\Omega)},
$$

for each $0 < p < \infty$, uniformly for $u \in C^0(\Omega, \mathcal{E})$ satisfying $Lu = 0$ in Ω.

REMARK III. The self-adjointness hypothesis can be relaxed to

$$L - L^* \in \mathrm{Diff}_1(\mathcal{E}, \mathcal{E}), \tag{3.56}$$

i.e., L is a first-degree perturbation of a formally self-adjoint operator, given that L^* also satisfies (2.70). Note that this together with the strong ellipticity assumption imply that $L : H_0^{1,2}(D, \mathcal{E}) \to H^{-1,2}(D, \mathcal{E})$ is Fredholm with index zero. Then, much as before, the non-singularity hypothesis allows us to infer that this operator is invertible for any Lipschitz domain $D \subseteq \mathcal{M}$ (by duality, a similar conclusion applies to L^* also). This suffices to carry out the proof of the Theorem 3.1 along the same lines, with natural modifications (note that, since L has a symmetric higher order part, (3.30) continues to hold).

REMARK IV. If the non-singularity hypothesis on L is dropped, one can show that the L^2-Dirichlet problem is Fredholm-solvable.

REMARK V. Sobolev type estimates in the spirit of (3.6) and (3.8) can also be derived for $p \neq 2$ by appealing to the last remark in §2. More specifically, for $2 - \varepsilon < p < 2 + \varepsilon$ we have

$$f \in L^p(\partial\Omega, \mathcal{E}) \Longrightarrow u \in B^{p,p^*}_{1/p}(\Omega, \mathcal{E}) \quad \text{and} \quad f \in H^{1,p}(\partial\Omega, \mathcal{E}) \Longrightarrow u \in B^{p,p^*}_{1+1/p}(\Omega, \mathcal{E})$$

(recall that $p^* = \max\{p, 2\}$). Naturally accompanying estimates are valid in each case.

REMARK VI. All the aforementioned results continue to hold if $\mathcal{M}$ is a compact manifold with boundary since, in this case, $\mathcal{M}$ can be embedded into a compact, boundaryless manifold of the same dimension.

Finally, elaborating on (3.55) above, we present a more general version of that square-function estimate.

COROLLARY 3.2. *Let $\mathcal{E} \to \mathcal{M}$ be as before. Also, consider a strongly elliptic, formally self-adjoint second-order differential operator L acting on sections of $\mathcal{E}$, whose coefficients satisfy (3.2) locally. Let Ω be a Lipschitz subdomain of $\mathcal{M}$ and suppose that $u \in C^0_{\mathrm{loc}}(\Omega, \mathcal{E})$ satisfies $Lu \in L^2(\Omega, \mathcal{E})$ and $\mathcal{N}(u) \in L^2(\partial\Omega)$.*

Then $u \in H^{1/2,2}(\Omega, \mathcal{E})$ and, in fact,

$$\iint_\Omega |\nabla u(x)|^2 \, dist\,(x, \partial\Omega)\, d\,Vol \leq C\|\mathcal{N}(u)\|^2_{L^2(\partial\Omega)} + C\|Lu\|^2_{L^2(\Omega,\mathcal{E})} \tag{3.57}$$

for some constant $C = C(L, \Omega) > 0$.

PROOF. Let $\lambda > 0$ be large enough so that $L - \lambda$ satisfies the non-singularity hypothesis relative to any Lipschitz subdomain of $\mathcal{M}$. In particular, $L - \lambda$ maps $H^{1,2}(\mathcal{M}, \mathcal{E})$ isomorphically onto $H^{-1,2}(\mathcal{M}, \mathcal{E})$. Set $f := Lu \in L^2(\Omega, \mathcal{E})$ and denote by Π_λ the Newtonian potential operator associated to $L - \lambda$, i.e. Π_λ is the integral operator in Ω whose kernel is the Schwartz kernel of $(L - \lambda)^{-1}$. By assumptions, $v := u - \Pi_\lambda(f - \lambda u)$ satisfies

$$\begin{cases} (L - \lambda)v = 0 \text{ in } \Omega, \\ v \in \{w \in C^0_{\text{loc}}(\Omega, \mathcal{E}); \, \mathcal{N}(w) \in L^2(\partial\Omega)\} + H^{2,2}(\Omega, \mathcal{E}). \end{cases} \tag{3.58}$$

Then (3.57) follows from (an adaptation of the corresponding proof in Theorem 3.1 for) the uniqueness in the L^2-Dirichlet problem for the operator $L - \lambda$ in Ω and (3.55). ■

CHAPTER 4

The Dirichlet Problem for the Hodge Laplacian and Related Operators

We retain from §§1–3 the hypotheses on the manifold $\mathcal{M}$ and, as much as possible, preserve notation introduced there. In addition, we will find it convenient to assume throughout the rest of this monograph that $\mathcal{M}$ is oriented. For many of our results, the hypothesis that $\mathcal{M}$ is orientable can be dropped, but for simplicity of presentation we will not discuss the nonorientable case. Note that no assumptions are made on the curvature.

Sections of $\Lambda^l T\mathcal{M}$, the l-th exterior power of the tangent bundle, consist of l-differential forms. If $(x_1, ..., x_m)$ are local coordinates in an arbitrary coordinate patch U on $\mathcal{M}$ and $u \in \Lambda^l T\mathcal{M}$ then $u|_U = \sum_{|I|=l} u_I \, dx^I$, where the sum is performed over ordered l-tuples $I = (i_1, ..., i_l)$, $1 \leq i_1 < i_2 < \cdots < i_l \leq m$, and, for each such I, $dx^I := dx_{i_1} \wedge \cdots \wedge dx_{i_l}$. Here, of course, the wedge stands for the usual exterior product of forms, while $|I|$ denotes the cardinality of I.

The Hermitian structure in the fibers on $T\mathcal{M}$ extends naturally to $T^*\mathcal{M}$ by setting $\langle dx_j, dx_k \rangle_x := g^{jk}(x)$. The latter further induces a Hermitian structure on $\Lambda^l T\mathcal{M}$ by selecting $\{\omega^I\}_{|I|=l}$ to be an orthonormal frame in $\Lambda^l T\mathcal{M}$ provided $\{\omega_j\}_{1 \leq j \leq m}$ is an orthonormal frame in $T^*\mathcal{M}$ (locally). Note that

$$\langle dx^I, dx^J \rangle_x = \det\left((g^{ij}(x))_{i \in I, j \in J} \right). \tag{4.1}$$

Next, recall that the (exterior) derivative operator $d \in \mathrm{Diff}_1(\Lambda^l T\mathcal{M}, \Lambda^{l+1} T\mathcal{M})$ is given by

$$d = \sum_{j=1}^{m} \tfrac{\partial}{\partial x_j} dx_j \wedge \cdot \quad \text{in } U. \tag{4.2}$$

Its formal adjoint (with respect to the metric described above), the co-differential operator δ, acts on a differential form $u \in \Lambda^l T\mathcal{M}$, $u|_U = \sum_{|I|=l} u_I \, dx^I$, by $\delta u = \sum_{|I|=l-1} (\delta u)_I \, dx^I$ where, for each I,

$$(\delta u)_I := -\sum_{|J|=l} \sum_{j,k} g^{jk} \varepsilon^J_{jI} \frac{\partial u_J}{\partial x_k} + \sum_{|K|=l-1} \sum_{j,k,r,s} \varepsilon^{rK}_{jI} g^{jk} \Gamma^s_{rk} u_{sK} \quad \text{in } U. \tag{4.3}$$

Here, for $1 \leq r, j, k \leq m$,

$$\Gamma^r_{jk} := \tfrac{1}{2} \sum_s g^{rs} \left(\frac{\partial g_{sj}}{\partial x_k} + \frac{\partial g_{sk}}{\partial x_j} - \frac{\partial g_{jk}}{\partial x_s} \right) \tag{4.4}$$

are the Christoffel symbols. Also, for any two ordered arrays J, K, the generalized Kronecker symbol ε^J_K is given by

$$\varepsilon^J_K := \begin{cases} \det\left((\delta_{j,k})_{j\in J, k\in K}\right), & \text{if } |J| = |K|, \\ 0, & \text{otherwise,} \end{cases} \tag{4.5}$$

where $\delta_{j,k} := 1$ if $j = k$, and zero if $j \neq k$.

We will make considerable use of the Hodge star operator, which can be characterized as the unique vector bundle morphism $* : \Lambda^l T\mathcal{M} \to \Lambda^{m-l} T\mathcal{M}$ such that

$$u \wedge (*u) = |u|^2 \, d\mathrm{Vol}.$$

Here we regard dVol as an m-form on $\mathcal{M}$, making use of the orientation we assume $\mathcal{M}$ has. We define the interior product between a 1-form α and an l-form u by setting

$$\alpha \vee u := (-1)^{(l-1)m} * (\alpha \wedge *u). \tag{4.6}$$

Thus, as it is well known, for $x \in \mathcal{M}$, $\xi \in T^*_x\mathcal{M} \setminus 0$,

$$\sigma(d)(x,\xi)u = i\,\xi \wedge u, \quad \text{and} \quad \sigma(\delta)(x,\xi)u = -i\,\xi \vee u. \tag{4.7}$$

In particular, if Ω is a Lipschitz subdomain of $\mathcal{M}$ with outward unit conormal ν defined $d\sigma$-a.e. on $\partial\Omega$, and $u \in C^1(\bar\Omega, \Lambda^l T\mathcal{M})$, $v \in C^1(\bar\Omega, \Lambda^{l+1} T\mathcal{M})$, then

$$\iint_\Omega \langle du, v\rangle \, d\mathrm{Vol} - \iint_\Omega \langle u, \delta v\rangle \, d\mathrm{Vol} = \int_{\partial\Omega} \langle \nu \wedge u, v\rangle \, d\sigma = \int_{\partial\Omega} \langle u, \nu \vee v\rangle \, d\sigma. \tag{4.8}$$

For further reference as well as for the convenience of the reader, some basic, elementary properties of these objects are summarized in the following lemma.

LEMMA 4.1. *For arbitrary one-forms α, β, and any l-form u, $(m-l)$-form v, and $(l+1)$-form w, the following are true:*

(1) $**u = (-1)^{l(m-l)}\, u$;
(2) $\langle u, *v\rangle = (-1)^{l(m-l)} \langle *u, v\rangle$ *and* $\langle *u, *v\rangle = \langle u, v\rangle$;
(3) $\alpha \wedge (\alpha \wedge u) = 0$ *and* $\alpha \vee (\alpha \vee u) = 0$;
(4) $\alpha \wedge (\beta \vee u) + \beta \vee (\alpha \wedge u) = \langle \alpha, \beta\rangle u$;
(5) $\langle \alpha \wedge u, w\rangle = \langle u, \alpha \vee w\rangle$;
(6) $*(\alpha \wedge u) = (-1)^l \alpha \vee (*u)$ *and* $*(\alpha \vee u) = (-1)^{l-1} \alpha \wedge (*u)$.

Moreover, if α is normalized such that $\langle \alpha, \alpha\rangle = 1$, then also:

(7) $u = \alpha \wedge (\alpha \vee u) + \alpha \vee (\alpha \wedge u)$;
(8) $|\alpha \wedge (\alpha \vee u)| = |\alpha \vee u|$ *and* $|\alpha \vee (\alpha \wedge u)| = |\alpha \wedge u|$.

Finally,

(9) $d\,d = 0$, $\delta\,\delta = 0$;
(10) $\delta = (-1)^{m(l+1)+1} * d*$ *and* $*\delta = (-1)^l d*$, $\delta* = (-1)^{l+1} * d$ *on l-forms.*

Next we discuss the Hodge Laplacian

$$\Delta := -(d\delta + \delta d). \tag{4.9}$$

Note that in order to make sense of Δ in (4.9) as an element in $\mathrm{Diff}_2(\Lambda^l T\mathcal{M}, \Lambda^l T\mathcal{M})$, $l = 0, 1, ..., m$, with coefficients satisfying (3.2) it suffices to assume that

(4.10) the Riemannian metric tensor g has coefficients in $H^{2,r}$, $r > m$, which we shall do from now on unless otherwise specified. This is because for an l-differential form $u \in C^2(\mathcal{M}, \Lambda^l T\mathcal{M})$, locally written as $u|_U = \sum_{|I|=l} u_I \, dx^I$, we have in U

$$\begin{aligned}
\Delta u = & \sum_{|I|=l} \sum_{j,k} g^{jk} \frac{\partial^2 u_I}{\partial x_j \partial x_k} \, dx^I - \sum_{|I|=l} \sum_{|M|=l} \sum_{|K|=l} \sum_{j,k,r,s,t} \varepsilon^{rK}_{jI} \varepsilon^{sK}_{tM} g^{jk} \Gamma^s_{rk} \frac{\partial u_M}{\partial x_t} \, dx^I \\
& - \sum_{|I|=l} \sum_{|J|=l-1} \sum_{|K|=l-1} \sum_{j,k,r,s,t} \varepsilon^{I}_{tJ} \varepsilon^{rK}_{jJ} \, g^{jk} \Gamma^s_{rk} \frac{\partial u_{sK}}{\partial x_t} \, dx^I \\
& - \sum_{|I|=l} \sum_{|J|=l-1} \sum_{|K|=l-1} \sum_{j,k,r,s,t} \varepsilon^{I}_{tJ} \varepsilon^{rK}_{jJ} \, u_{sK} \frac{\partial (g^{jk} \Gamma^s_{rk})}{\partial x_t} \, dx^I .
\end{aligned} \tag{4.11}$$

Let us point out that for zero-forms, i.e., scalar-valued functions on $\mathcal{M}$, the zero order terms in Δ are absent due to obvious degree considerations. Consequently, the corresponding theory for harmonic (scalar-valued) functions can be worked out under the weaker assumption $g \in C^1$; cf. [MT1].

Recall that a form u is called *harmonic* if $\Delta u = 0$ in the distributional sense. The next theorem treats the Dirichlet problem for harmonic forms of arbitrary degree in Lipschitz domains on manifolds.

THEOREM 4.2. *Let $\mathcal{M}$ be as above and consider some arbitrary Lipschitz domain $\Omega \subseteq \mathcal{M}$ with non-empty boundary. Then there exists $\varepsilon = \varepsilon(\Omega) > 0$ such that, for any form $f \in L^p(\partial\Omega, \Lambda^l T\mathcal{M})$, where $2 - \varepsilon < p < 2 + \varepsilon$ and $0 \leq l \leq m$, the Dirichlet boundary problem*

$$\begin{cases} u \in C^0(\Omega, \Lambda^l T\mathcal{M}), \\ \Delta u = 0 \text{ in } \Omega, \\ \mathcal{N}(u) \in L^p(\partial\Omega), \\ u|_{\partial\Omega} = f \text{ on } \partial\Omega, \end{cases} \tag{4.12}$$

has a unique solution. This solution belongs to $C^{1+\mu}(\Omega, \Lambda^l T\mathcal{M})$ for some $\mu > 0$ and satisfies

$$\|\mathcal{N} u\|_{L^p(\partial\Omega)} \leq C \|f\|_{L^p(\partial\Omega, \Lambda^l T\mathcal{M})} \tag{4.13}$$

for some positive constant C, independent of f.

Also, $u \in H^{1/2,2}(\Omega, \Lambda^l T\mathcal{M})$ for any $f \in L^2(\partial\Omega, \Lambda^l T\mathcal{M})$ and

$$\|u\|_{H^{1/2,2}(\Omega, \Lambda^l T\mathcal{M})} \leq C \|f\|_{L^2(\partial\Omega, \Lambda^l T\mathcal{M})} \tag{4.14}$$

for some $C > 0$ independent of f.

Further, if f actually belongs to $H^{1,p}(\partial\Omega, \Lambda^l T\mathcal{M})$, $|p - 2| \leq \varepsilon$, then $\mathcal{N}(\nabla u) \in L^p(\partial\Omega)$ and

$$\|\mathcal{N}(\nabla u)\|_{L^p(\partial\Omega)} \leq C \|f\|_{H^{1,p}(\partial\Omega, \Lambda^l T\mathcal{M})} \tag{4.15}$$

for some $C = C(\Omega) > 0$. Moreover, $u \in H^{3/2,2}(\Omega, \Lambda^l T\mathcal{M})$ for any f belonging to $H^{1,2}(\partial\Omega, \Lambda^l T\mathcal{M})$ and

$$\|u\|_{H^{3/2,2}(\Omega, \Lambda^l T\mathcal{M})} \leq C \|f\|_{H^{1,2}(\partial\Omega, \Lambda^l T\mathcal{M})} \tag{4.16}$$

for some $C > 0$ independent of f.

The case of zero-degree forms in the "flat" Euclidean setting is well-known and goes back to B. Dahlberg [Da], E. Fabes, M. Jodeit and N. Rivière [FJR], D. Jerison and C. Kenig [JK1], and G. Verchota [Ve1]. The case $l = 0$ has been recently extended to Lipschitz domains on manifolds by M. Mitrea and M. Taylor [MT1]. Variants of this Dirichlet problem for general $0 \leq l \leq m$ in the context of smooth domains, smooth metric structures and smooth boundary data have been previously considered by G. F. D. Duff and D. C. Spencer [DS1], [DS2], [Du3], [Sp], C. B. Morrey and J. Eells [ME1], [ME2], and G. Schwarz [Sc].

The proof of Theorem 4.2 rests on our results in §§1–3 and a deep unique continuation theorem due to N. Aronszajn, K. Krzywicki and J. Szarski [AKS]. Since we shall use the latter several times in the sequel, we record the precise statement below.

THEOREM 4.3 [AKS]. *Let $\mathcal{M}$ be a connected Riemannian manifold with a Lipschitz metric tensor and let $u \in H^{1,2}_{\rm loc}(\mathcal{M}, \Lambda^l T\mathcal{M})$ be such that*

$$|du|^2 + |\delta u|^2 \leq C_K |u|^2 \tag{4.17}$$

pointwise a.e. on each compact subset $K \subseteq \mathcal{M}$. If u vanishes on an open subset of $\mathcal{M}$ then $u \equiv 0$ on $\mathcal{M}$.

PROOF OF THEOREM 4.2. Consider $V \in L^\infty(\mathcal{M})$ a positive scalar-valued function so that $V > 0$ in some open subset of $\mathcal{M}$ and $\operatorname{supp} V \cap \bar{\Omega} = \varnothing$. Then the second order differential operator $L := \Delta - V$ is self-adjoint and strongly elliptic, and we claim that it also satisfies the nonsingularity hypothesis (3.3) relative to any Lipschitz subdomain D of $\mathcal{M}$.

Indeed, let $u \in H^{1,2}_0(D, \Lambda^l T\mathcal{M})$ be such that $Lu = 0$ in D. Then, (4.9) and the integration by parts formula (4.8) give $\iint_D \left(|du|^2 + |\delta u|^2 + V|u|^2\right) d\mathrm{Vol} = 0$. Consequently,

$$u = 0 \text{ on } \operatorname{supp} V, \text{ and } du = 0, \ \delta u = 0 \text{ in } D. \tag{4.18}$$

Therefore, if $u \mapsto \tilde{u}$ denotes the extension by zero outside D, it follows that $\tilde{u}$ belongs to $H^{1,2}(\mathcal{M}, \Lambda^l T\mathcal{M})$ and is a harmonic field, i.e.,

$$d\tilde{u} = 0 \text{ and } \delta\tilde{u} = 0 \text{ in } \mathcal{M}. \tag{4.19}$$

Since it vanishes identically in an open set, we infer from Theorem 4.3 that $\tilde{u}$ vanishes identically in $\mathcal{M}$, i.e. $u = 0$ in D. Thus, everything is seen from Theorem 3.1 (cf. also the remark preceding its proof). ■

Before going any further, we pause for a moment in order to record a few remarks. Firstly, the Dirichlet boundary condition in (4.12) amounts to prescribing both $u_{\rm tan}$ in $L^p_{\rm tan}(\partial\Omega, \Lambda^l T\mathcal{M})$ and $u_{\rm nor}$ in $L^p_{\rm nor}(\partial\Omega, \Lambda^l T\mathcal{M})$. Here

$$u_{\rm tan} := \nu \vee (\nu \wedge u|_{\partial\Omega}), \quad u_{\rm nor} := \nu \wedge (\nu \vee u|_{\partial\Omega}) \tag{4.20}$$

stand, respectively, for the *tangential* and *normal* components of u on $\partial\Omega$, and

$$\begin{aligned} L^p_{\rm tan}(\partial\Omega, \Lambda^l T\mathcal{M}) &:= \{v \in L^p(\partial\Omega, \Lambda^l T\mathcal{M});\ v_{\rm nor} = 0\}, \\ L^p_{\rm nor}(\partial\Omega, \Lambda^l T\mathcal{M}) &:= \{v \in L^p(\partial\Omega, \Lambda^l T\mathcal{M});\ v_{\rm tan} = 0\}, \end{aligned} \tag{4.21}$$

$1 < p < \infty$, $l = 0, 1, \dots, m$. A measurable section $f : \partial\Omega \to \Lambda^l T\mathcal{M}$ is called *tangential* if $\nu \vee f = 0$ a.e. on $\partial\Omega$, and *normal* if $\nu \wedge f = 0$ on $\partial\Omega$. In passing, let us note that, for any u, $u_{\tan}$ is tangential, $u_{\rm nor}$ is normal, $u = u_{\tan} + u_{\rm nor}$ and $\langle u_{\tan}, u_{\rm nor}\rangle = 0$. In particular, $|u|^2 = |u_{\tan}|^2 + |u_{\rm nor}|^2$. Finally, the pair of boundary conditions $u_{\tan}$, $u_{\rm nor}$ = given on $\partial\Omega$, is equivalent to prescribing

$$\nu \wedge u|_{\partial\Omega} \ \text{ in } \ L^p_{\rm nor}(\partial\Omega, \Lambda^{l+1}T\mathcal{M}) \tag{4.22}$$

together with

$$\nu \vee u|_{\partial\Omega} \ \text{ in } \ L^p_{\tan}(\partial\Omega, \Lambda^{l-1}T\mathcal{M}). \tag{4.23}$$

Other pairs of boundary conditions (involving also derivatives of u) will be considered in subsequent chapters.

Secondly, in the case treated in Theorem 4.2, the first-order differential operator $P \in \mathrm{Diff}_1(\Lambda^l T\mathcal{M}, \Lambda^l T\mathcal{M})$ appearing in the proof of Theorem 3.1 can be taken to be $P(x, D) = d(w \vee \cdot) + w \vee d$, for some smooth $w \in T^*\mathcal{M}$ satisfying $\langle w, \nu\rangle \geq C > 0$ on $\partial\Omega$. Consequently, the solution u of (4.12) can be looked for in the form

$$\begin{aligned} u(x) = & \int_{\partial\Omega} \langle d_y(w(y) \vee \Gamma(x,y)), h(y)\rangle_y \, d\sigma(y) \\ & + \int_{\partial\Omega} \langle d_y\Gamma(x,y), w(y) \wedge h(y)\rangle_y \, d\sigma(y) \\ & + \int_{\partial\Omega} \langle \Gamma(x,y), \tilde{h}(y)\rangle_y \, d\sigma(y), \end{aligned} \tag{4.24}$$

$x \in \Omega$, for suitable $h, \tilde{h} \in L^p(\partial\Omega, \Lambda^l T\mathcal{M})$, where $\Gamma(x,y)$ is the Schwartz kernel of $(\Delta - V)^{-1}$ acting on l-forms (cf. also the discussion in §6).

Thirdly, it is important to point out that it is straightforward to adapt the results above to the case of vector bundle-valued differential forms (in an appropriate smoothness context). The main points are that the (generalized) Laplacian associated with some connection on the vector bundle is strongly elliptic and self-adjoint, and that Theorem 4.3 continues to hold in this context.

The last observation we wish to make at this stage is that the Besov regularity results stated in Remark IV of §3 continue to hold in the present setting. They represent natural extensions of (4.14) and (4.16) to the case when $p \neq 2$.

Returning now to the main line of discussion in this chapter, we note that the same argument used to prove Theorem 4.2 yields a more general result. To state it, we say that a first-order differential operator $D : \mathcal{E} \to \mathcal{F}$ has the unique continuation property (henceforth denoted UCP) if $\Omega \subset \mathcal{M}$ Lipschitz, $u \in H^{1,2}_0(\Omega; \mathcal{E}, \mathcal{F})$ and $Du = 0$ in Ω imply $u = 0$.

THEOREM 4.4. *Let $\mathcal{E}, \mathcal{F} \to \mathcal{M}$ be as in §§1 – 3 and consider $D : \mathcal{E} \to \mathcal{F}$ an elliptic first-order differential operator satisfying the UCP. It is assumed that the coefficients of $L := D^*D$ satisfy the regularity assumptions (3.2).*

Then, for any $\Omega \subset \mathcal{M}$ Lipschitz domain there exists $\varepsilon = \varepsilon(\Omega, \mathcal{E}, \mathcal{F}, L) > 0$ such that, for any section $f \in L^p(\partial\Omega, \mathcal{E})$, where $2 - \varepsilon < p < 2 + \varepsilon$, the Dirichlet boundary *problem*

$$
(4.25) \qquad \begin{cases} u \in C^0(\Omega, \mathcal{E}), \\ Lu = 0 \text{ in } \Omega, \\ \mathcal{N}(u) \in L^p(\partial\Omega), \\ u|_{\partial\Omega} = f \text{ on } \partial\Omega, \end{cases}
$$

has a unique solution. Furthermore, this solution satisfies regularity properties similar to the ones described in Theorem 4.2 (cf. also the remarks following its proof).

The case dealt with in Theorem 4.2 occurs, e.g. for $D := (d, \delta)$. Lamé type operators, i.e. of the form $L := \Delta + \delta\phi d$ for positive, operator-valued functions $\phi \in C^{1+\gamma}(\mathcal{M}, \mathrm{Hom}(\Lambda^l T\mathcal{M}, \Lambda^l T\mathcal{M}))$, $\gamma > 0$, can also be obtained in a similar fashion (in fact, certain lower order terms may also be considered).

Another example of interest is that of the symmetric exterior differentiation operator; see, e.g., [Sh; p. 88-93] for a discussion of ellipticity and UCP as far as this operator is concerned. Finally, we note that (under appropriate smoothness assumptions on the coefficients), by [Ar], the operator D has UCP whenever L has a scalar principal part.

CHAPTER 5

Natural Boundary Problems for the Hodge Laplacian in Lipschitz Domains

In this chapter we undertake a systematic study of natural boundary problems for the Hodge-Laplacian. Several of those have a mixed character, in as much as (parts of) both u and derivatives of u are prescribed on the boundary. As before, let $\mathcal{M}$ be a smooth, connected, compact, boundaryless manifold of real dimension m, equipped with a Riemannian metric

$$g \in H^{2,r}(\mathcal{M}, \mathrm{Hom}\,(T\mathcal{M} \otimes T\mathcal{M}, \mathbb{R})), \quad r > m = \dim \mathcal{M}.$$

Fix Ω an arbitrary Lipschitz domain in $\mathcal{M}$ with nonempty boundary. An l-form $f \in L^p_{\tan}(\partial\Omega, \Lambda^l T\mathcal{M})$, $1 \leq p \leq \infty$, is said to have its *boundary (exterior) co-derivative* in L^p if there exists an $(l-1)$-form in $L^p(\partial\Omega, \Lambda^{l-1} T\mathcal{M})$, which we denote by $\delta_\partial f$, so that

$$(5.1) \qquad \int_{\partial\Omega} \langle d\psi, f \rangle \, d\sigma = \int_{\partial\Omega} \langle \psi, \delta_\partial f \rangle \, d\sigma \quad \text{for any } \psi \in C^1(\mathcal{M}, \Lambda^{l-1} T\mathcal{M}).$$

We set

$$(5.2) \qquad L^{p,\delta}_{\tan}(\partial\Omega, \Lambda^l T\mathcal{M}) := \left\{ f \in L^p_{\tan}(\partial\Omega, \Lambda^l T\mathcal{M});\ \delta_\partial f \in L^p(\partial\Omega, \Lambda^{l-1} T\mathcal{M}) \right\},$$

and equip it with the natural norm

$$(5.3) \qquad \|f\|_{L^{p,\delta}_{\tan}(\partial\Omega, \Lambda^l T\mathcal{M})} := \|f\|_{L^p(\partial\Omega, \Lambda^l T\mathcal{M})} + \|\delta_\partial f\|_{L^p(\partial\Omega, \Lambda^{l-1} T\mathcal{M})}.$$

It is immediate that δ_∂ is a local operator, i.e. $\mathrm{supp}\,(\delta_\partial f) \subseteq \mathrm{supp}\, f$ for any form f in $L^{p,\delta}_{\tan}(\partial\Omega, \Lambda^l T\mathcal{M})$ and, hence, in the smooth context, δ_∂ is a first order differential operator. However, the coefficients would involve *second order* derivatives of the transition functions between local charts in the manifold $\partial\Omega$ and, hence, such a description is no longer possible in the Lipschitz case. It is precisely for circumventing this problem that we resort to the distributional definition above. In the Euclidean setting, this definition was first introduced in [Mi1], [Mi2] and subsequently utilized in [JM], [MD1], [MM1].

It is not difficult to check that

$$(5.4) \qquad \delta_\partial \delta_\partial f = 0 \text{ and } \nu \vee \delta_\partial f = 0 \text{ on } \partial\Omega \text{ for any } f \in L^{p,\delta}_{\tan}(\partial\Omega, \Lambda^l T\mathcal{M}),$$

where recall that ν stands for the outward unit conormal to $\partial\Omega$. In particular, introducing the (closed) subspace $L^{p,0}_{\tan}(\partial\Omega, \Lambda^l T\mathcal{M})$ of $L^{p,\delta}_{\tan}(\partial\Omega, \Lambda^l T\mathcal{M})$ by

$$(5.5) \qquad L^{p,0}_{\tan}(\partial\Omega, \Lambda^l T\mathcal{M}) := \left\{ f \in L^{p,\delta}_{\tan}(\partial\Omega, \Lambda^l T\mathcal{M});\ \delta_\partial f = 0 \right\},$$

then

$$\delta_\partial : L^{p,\delta}_{\tan}(\partial\Omega, \Lambda^l T\mathcal{M}) \longrightarrow L^{p,0}_{\tan}(\partial\Omega, \Lambda^{l-1} T\mathcal{M}) \tag{5.6}$$

is well defined and bounded.

In order to state our first result, for $1 < p < \infty$ and $0 \le l \le m$ we introduce

$$\mathcal{H}^{l,p}_\vee(\Omega) := \{u \in C^1(\Omega, \Lambda^l T\mathcal{M});\ \mathcal{N}(u) \in L^p(\partial\Omega),\ du = 0,\ \delta u = 0 \text{ in } \Omega,\ \nu \vee u\big|_{\partial\Omega} = 0\}. \tag{5.7}$$

As we shall prove in §11, the space $\mathcal{H}^{l,p}_\vee(\Omega)$ is actually *independent* of p provided $2 - \varepsilon < p < 2 + \varepsilon$ for some $\varepsilon = \varepsilon(\Omega) > 0$. In such cases, we shall occasionally drop the subscript p and simply write $\mathcal{H}^l_\vee(\Omega)$ in place of $\mathcal{H}^{l,p}_\vee(\Omega)$.

THEOREM 5.1. *Let Ω be a Lipschitz domain in $\mathcal{M}$ and, for each $l \in \{0, 1, ..., m\}$, consider the boundary value problem*

$$(BVP1)_l \begin{cases} u \in C^1(\Omega, \Lambda^l T\mathcal{M}), \\ \Delta u = 0 \text{ in } \Omega, \\ \mathcal{N}(u), \mathcal{N}(du) \in L^p(\partial\Omega), \\ \nu \vee u\big|_{\partial\Omega} = f \in L^p_{\tan}(\partial\Omega, \Lambda^{l-1} T\mathcal{M}), \\ \nu \vee (du)\big|_{\partial\Omega} = g \in L^p_{\tan}(\partial\Omega, \Lambda^l T\mathcal{M}). \end{cases}$$

Then there exists $\varepsilon = \varepsilon(\Omega)$ such that, if $2 - \varepsilon < p < 2 + \varepsilon$, we have the following:

1. *A solution of $(BVP1)_l$ exists if and only if g satisfies the compatibility condition*

$$g \in \{v\big|_{\partial\Omega};\ v \in \mathcal{H}^l_\vee(\Omega)\}^\circ; \tag{5.8}$$

here "$\{...\}^\circ$" refers to the annihilator of $\{...\}$ ($\subseteq L^q(\partial\Omega, \Lambda^l T\mathcal{M})$, $1/p + 1/q = 1$) in $L^p_{\tan}(\partial\Omega, \Lambda^l T\mathcal{M})$.

2. *The dimension of the space of null-solutions for the homogeneous problem is $b_l(\Omega)$, the l-th Betti number of Ω, and, in fact, this space coincides precisely with $\mathcal{H}^l_\vee(\Omega)$. In particular, the boundary data determine du and δu uniquely and*

$$\|\mathcal{N}(du)\|_{L^p(\partial\Omega)} \le C\left(\|g\|_{L^p(\partial\Omega, \Lambda^l T\mathcal{M})} + \|f\|_{L^p(\partial\Omega, \Lambda^{l-1} T\mathcal{M})}\right) \tag{5.9}$$

for some positive constant C, independent of f, g.

Also, for $p = 2$, every solution belongs to $H^{1/2,2}(\Omega, \Lambda^l T\mathcal{M})$ (more generally, for $2 - \varepsilon < p < 2 + \varepsilon$, each solution belongs to $B^{p,p^*}_{1/p}(\Omega, \Lambda^l T\mathcal{M})$).

Moreover, if the compatibility condition (5.8) is satisfied, then the following regularity statements are valid:

3. *$\mathcal{N}(\delta u) \in L^p(\partial\Omega)$ if and only if $f \in L^{p,\delta}_{\tan}(\partial\Omega, \Lambda^{l-1} T\mathcal{M})$. Moreover, for $f \in L^{p,\delta}_{\tan}(\partial\Omega, \Lambda^{l-1} T\mathcal{M})$, there holds*

$$\|\mathcal{N}(\delta u)\|_{L^p(\partial\Omega)} \le C\left(\|f\|_{L^{p,\delta}_{\tan}(\partial\Omega, \Lambda^{l-1} T\mathcal{M})} + \|g\|_{L^p(\partial\Omega, \Lambda^l T\mathcal{M})}\right). \tag{5.10}$$

4. *$\mathcal{N}(\delta du) \in L^p(\partial\Omega)$ if and only if $g \in L^{p,\delta}_{\tan}(\partial\Omega, \Lambda^l T\mathcal{M})$. In this case*

$$\|\mathcal{N}(\delta du)\|_{L^p(\partial\Omega)} \le C\|g\|_{L^{p,\delta}_{\tan}(\partial\Omega, \Lambda^l T\mathcal{M})}. \tag{5.11}$$

5. *$du = 0$ in Ω if and only if $g = 0$. Furthermore, when $g = 0$ we can prescribe periods and have genuine uniqueness. More precisely, for any $\beta_j \in \mathbb{R}$, $j = 1, ..., b_l(\Omega)$, there exists a unique solution of*

$$(BVP2)_l \begin{cases} u \in C^1(\Omega, \Lambda^l T\mathcal{M}), \\ \Delta u = 0 \text{ in } \Omega, \\ du = 0 \text{ in } \Omega, \\ \mathcal{N}(u) \in L^p(\partial\Omega), \\ \nu \vee u\big|_{\partial\Omega} = f \in L^p_{\tan}(\partial\Omega, \Lambda^{l-1}T\mathcal{M}), \\ \int_{\gamma_j} \iota^* u = \beta_j, \ j = 1, ..., b_l(\Omega), \end{cases}$$

where $\{[\gamma_j]\}_{j=1,...,b_l(\Omega)}$ is a basis of $H^l_{\rm sing}(\Omega; \mathbb{R})$, the l-th singular homology group of Ω over the reals and, for all j's, $\iota : \gamma_j \hookrightarrow \Omega$ is the inclusion.

6. *$\delta du = 0$ in Ω if and only if $g \in L^{p,0}_{\tan}(\partial\Omega, \Lambda^l T\mathcal{M})$. In particular, for g belonging to $L^{p,0}_{\tan}(\partial\Omega, \Lambda^l T\mathcal{M})$ the problem $(BVP1)_l$ becomes*

$$(BVP3)_l \begin{cases} u \in C^1(\Omega, \Lambda^l T\mathcal{M}), \\ \Delta u = 0 \text{ in } \Omega, \\ \delta du = 0 \text{ in } \Omega, \\ \mathcal{N}(u), \mathcal{N}(du) \in L^p(\partial\Omega), \\ \nu \vee u\big|_{\partial\Omega} = f \in L^p_{\tan}(\partial\Omega, \Lambda^{l-1}T\mathcal{M}), \\ \nu \vee (du)\big|_{\partial\Omega} = g \in L^{p,0}_{\tan}(\partial\Omega, \Lambda^l T\mathcal{M}). \end{cases}$$

7. *Whenever $f \in L^{p,\delta}_{\tan}(\partial\Omega, \Lambda^{l-1}T\mathcal{M})$ we can prescribe $\nu \vee (\delta u)\big|_{\partial\Omega} = \delta_\partial f$ arbitrarily in $\delta_\partial L^{p,\delta}_{\tan}(\partial\Omega, \Lambda^{l-1}T\mathcal{M})$ in place of $\nu \vee u\big|_{\partial\Omega}$ in any of the problems $(BVP1)_l - (BVP3)_l$. Formulated as such, each of these problems has an infinite dimensional null space.*

8. *$\delta u = 0$ if and only if*

$$f \in L^{p,0}_{\tan}(\partial\Omega, \Lambda^{l-1}T\mathcal{M}) \cap \{\mathcal{H}^{l-1}_\vee(\Omega)\big|_{\partial\Omega}\}^\circ \text{ and } g \in L^{p,0}_{\tan}(\partial\Omega, \Lambda^l T\mathcal{M}). \tag{5.12}$$

In fact, if f and g are as above then $(BVP1)_l$ reduces to

$$(BVP4)_l \begin{cases} u \in C^1(\Omega, \Lambda^l T\mathcal{M}), \\ \Delta u = 0 \text{ in } \Omega, \\ \delta u = 0 \text{ in } \Omega, \\ \mathcal{N}(u), \mathcal{N}(du) \in L^p(\partial\Omega), \\ \nu \vee u\big|_{\partial\Omega} = f \in L^{p,0}_{\tan}(\partial\Omega, \Lambda^{l-1}T\mathcal{M}), \\ \nu \vee (du)\big|_{\partial\Omega} = g \in L^{p,0}_{\tan}(\partial\Omega, \Lambda^l T\mathcal{M}). \end{cases}$$

In particular, $g = 0$ forces $du = 0$ and we can prescribe periods, in which case $(BVP4)_l$ becomes

$$(BVP5)_l \begin{cases} u \in C^1(\Omega, \Lambda^l TM), \\ \delta u = 0 \text{ in } \Omega, \\ du = 0 \text{ in } \Omega, \\ \mathcal{N}(u) \in L^p(\partial\Omega), \\ \nu \vee u\big|_{\partial\Omega} = f \in L^{p,0}_{\tan}(\partial\Omega, \Lambda^{l-1}TM), \\ \int_{\gamma_j} \iota^* u = \beta_j, \ j = 1, ..., b_l(\Omega). \end{cases}$$

Formulated as such, the problem $(BVP5)_l$ has a solution if and only if $f \in \{\mathcal{H}^{l-1}_\vee(\Omega)\big|_{\partial\Omega}\}^\circ$ and the solution is unique.

9. *Finally, there hold analogous statements to (1)–(8) above for the dual of $(BVP1)_l$, that is*

$$(BVP6)_l \begin{cases} u \in C^1(\Omega, \Lambda^l TM), \\ \Delta u = 0 \text{ in } \Omega, \\ \mathcal{N}(u), \mathcal{N}(\delta u) \in L^p(\partial\Omega), \\ \nu \wedge u\big|_{\partial\Omega} = f \in L^p_{\mathrm{nor}}(\partial\Omega, \Lambda^{l+1}TM), \\ \nu \wedge (\delta u)\big|_{\partial\Omega} = g \in L^p_{\mathrm{nor}}(\partial\Omega, \Lambda^l TM). \end{cases}$$

In this case, we replace the space (5.7) by

$$\begin{aligned} \mathcal{H}^l_\wedge(\Omega) := *\mathcal{H}^{m-l}_\vee(\Omega) = \{u \in & C^1(\Omega, \Lambda^l TM); \mathcal{N}(u) \in L^p(\partial\Omega), \\ & du = 0, \ \delta u = 0 \text{ in } \Omega, \ \nu \wedge u\big|_{\partial\Omega} = 0\}. \end{aligned} \tag{5.13}$$

The boundary conditions in $(BVP1)_l$ are often called "absolute boundary conditions" and those in $(BVP6)_l$ are called "relative boundary conditions". It is both illuminating and rewarding to point out that $(BVP1)_l$ becomes the Dirichlet problem for the Laplacian (in slight disguise) for $l = m$ since, by Lemma 4.1,

$$u|_{\partial\Omega} = \nu \wedge (\nu \vee u|_{\partial\Omega}) = \nu \wedge f \in L^p(\partial\Omega, \Lambda^m TM) \cong L^p(\partial\Omega) d\mathrm{Vol},$$

whereas $L^p_{\tan}(\partial\Omega, \Lambda^m TM) = 0$ and $du = 0$. Furthermore, the problem above also contains its so-called regular version if, in addition, the boundary data f belongs to the Sobolev-type space $L^{p,\delta}_{\tan}(\partial\Omega, \Lambda^m TM)$. This is because the latter space can be naturally identified with $H^{1,2}(\partial\Omega) d\mathrm{Vol}$. In the same vein, when $l = 0$ $(BVP1)_l$ reduces precisely to the classical Neumann problem for the Laplacian acting on scalar-valued functions since, in this case, $\nu \vee u = 0$ while

$$\nu \vee (du)|_{\partial\Omega} = \langle \nu, du \rangle = \frac{\partial u}{\partial \nu}.$$

In the flat, Euclidean setting, these limiting cases are well understood from the work of B. Dahlberg [Da], E. Fabes, M. Jodeit and N. Riviére [FJR], D. Jerison and C. Kenig [JK1], G. Verchota [Ve1]. Actually, these papers deal only with domains with a trivial topology, a restriction which has been eliminated in [MD2].

Quite recently, such results have been extended to arbitrary Lipschitz domains in Riemannian manifolds by M. Mitrea and M. Taylor [MT1]. Also, in the flat, Euclidean case but for arbitrary degrees $l \in \{0, 1, ..., m\}$, similar problems have been addressed in the work of D. Mitrea and M. Mitrea [MD1], [MM1]. In the context of arbitrary domains in manifolds and for arbitrary degrees $0 \leq l \leq m$, the only reasonably well understood case is that when all structures involved are smooth. In this context, problems related to $(BVP1)_l$ have been treated by P. E. Conner [Co], G. F. D. Duff [Du4], P. R. Garabedian and D. C. Spencer [GS2], and G. Schwarz [Sc]. See also [Ta2] for a presentation. A notable exception is C. B. Morrey ([Mo1], [Mo3]) who uses variational methods and a priori regularity estimates; however, in his formulation, boundary traces are taken in a weak, distributional sense which actually affects the character of the problem.

The formulation of the problem $(BVP2)_l$, at least in a simpler geometric and analytic context, goes back to Hodge himself ([Hod1]). His pioneering work, which has drawn an enormous amount of consideration over the years, has been extended to smooth domains in manifolds by G. F. D. Duff, P. R. Garabedian, J. J. Kohn and D. C. Spencer in [DS1], [DS2], [GS1], [GS2], [Du4], [KS] and [Sp].

The families $(BVP1)_l$–$(BVP6)_l$ also encompass basic problems arising in static electromagnetism and fluid mechanics. Typically, this identification takes place at the level $l = 1$ when one identifies differential forms of degree one with ordinary vector fields. In particular, $(BVP5)_1$ contains an old problem of Lord Kelvin ([Kel]) regarding the determination of the motion of an incompressible, irrotational fluid occupying the domain Ω, given the normal velocity on the boundary and the circulation around the "handles". Results concerning the higher degree case of this problem have been announced by A. W. Tucker in his early work on this subject ([Tu]), while proofs of these conjectures have been subsequently supplied by G. F. D. Duff and D. C. Spencer in [Du1], [DS1], [DS2]. Once again, all these authors deal with the smooth case of the problem under discussion.

Let us also point out that the range $2 - \varepsilon < p < 2 + \varepsilon$ is optimal in the class of (all smooth manifolds and) all Lipschitz domains if one insists on either allowing arbitrary degrees $l \in \{0, 1, ..., m\}$, or staying away from the endpoints, i.e. for $1 \leq l \leq m - 1$. The sharp ranges for the extreme values $l = 0, m$ are well-known from the work of B. Dahlberg and C. Kenig [DK] for the flat Laplacian. See also the work of D. Mitrea, M. Mitrea and J. Pipher [MMP] for the three dimensional case, where counterexamples to such an extension to L^p, for $l = 1, 2$ and $p \neq 2$ have been constructed (note that this stands in sharp contrast with the scalar-valued case).

Our final remark concerns the version of (5.10) in Theorem 5.1 when $l = 1$. Translated in the language of vector fields (and with familiar notation) this reads

$$\Delta \vec{u} = 0 \text{ in } \Omega,\ \mathcal{N}(\vec{u}),\ \mathcal{N}(\operatorname{curl} \vec{u}) \in L^p(\partial\Omega) \Rightarrow \mathcal{N}(\operatorname{div} \vec{u}) \in L^p(\partial\Omega).$$

This is remarkable since, generally speaking, $|\vec{u}|$ and $|\operatorname{curl} \vec{u}|$ do not dominate $|\operatorname{div} \vec{u}|$ pointwise.

The second natural boundary problem makes the object of the next theorem. To formulate it, we need to discuss first certain spaces of regular normal forms on $\partial\Omega$. For $f \in L^p_{\text{nor}}(\partial\Omega, \Lambda^l T\mathcal{M})$, we define the distribution $d_\partial f$ by requiring that

$$\int_{\partial\Omega} \langle \delta\psi, f \rangle \, d\sigma = \int_{\partial\Omega} \langle \psi, d_\partial f \rangle \, d\sigma \quad \text{for any } \psi \in C^1(\mathcal{M}, \Lambda^{l+1} T\mathcal{M}). \tag{5.14}$$

Set

$$L^{p,d}_{\text{nor}}(\partial\Omega, \Lambda^l T\mathcal{M}) := \left\{ f \in L^p_{\text{nor}}(\partial\Omega, \Lambda^l T\mathcal{M});\ d_\partial f \in L^p(\partial\Omega, \Lambda^{l+1} T\mathcal{M}) \right\}, \tag{5.15}$$

equipped with the natural norm

$$\|f\|_{L^{p,d}_{\rm nor}(\partial\Omega,\Lambda^l T\mathcal{M})} := \|f\|_{L^p(\partial\Omega,\Lambda^l T\mathcal{M})} + \|d_\partial f\|_{L^p(\partial\Omega,\Lambda^{l+1} T\mathcal{M})}.$$

Analogously to (5.4), we have that

$$d_\partial d_\partial f = 0 \text{ and } \nu\wedge d_\partial f = 0 \text{ on } \partial\Omega \text{ for any } f\in L^{p,d}_{\rm nor}(\partial\Omega,\Lambda^l T\mathcal{M}), \tag{5.16}$$

so that, if we set

$$L^{p,0}_{\rm nor}(\partial\Omega,\Lambda^l T\mathcal{M}) := \left\{f\in L^{p,d}_{\rm nor}(\partial\Omega,\Lambda^l T\mathcal{M});\, d_\partial f = 0\right\}, \tag{5.17}$$

then

$$d_\partial L^{p,d}_{\rm nor}(\partial\Omega,\Lambda^l T\mathcal{M}) \subseteq L^{p,0}_{\rm nor}(\partial\Omega,\Lambda^{l+1} T\mathcal{M}). \tag{5.18}$$

Finally, for $1<p<\infty$, we introduce

$$\mathcal{H}^{l,p}(\Omega) := \{v\in C^1(\Omega,\Lambda^l T\mathcal{M});\, \mathcal{N}v\in L^p(\partial\Omega),\ dv=0,\ \delta v=0 \text{ in } \Omega\}. \tag{5.19}$$

Following K. Kodaira [Ko], we shall call the forms of class C^1 which are annihilated both by d and δ *harmonic fields.*

THEOREM 5.2. *Let Ω be a Lipschitz domain in $\mathcal{M}$ and, for each $0\le l\le m$, consider the boundary value problem*

$$(BVP7)_l \begin{cases} u\in C^1(\Omega,\Lambda^l T\mathcal{M}),\\ \Delta u = 0 \text{ in } \Omega,\\ \mathcal{N}(u),\mathcal{N}(du),\mathcal{N}(\delta u)\in L^p(\partial\Omega),\\ \nu\vee(du)|_{\partial\Omega} = f\in L^p_{\rm tan}(\partial\Omega,\Lambda^l T\mathcal{M}),\\ \nu\wedge(\delta u)|_{\partial\Omega} = g\in L^p_{\rm nor}(\partial\Omega,\Lambda^l T\mathcal{M}).\end{cases}$$

Then there exists $\varepsilon=\varepsilon(\Omega)>0$ such that the following hold whenever $2-\varepsilon<p<2+\varepsilon$:

1. *A solution of $(BVP7)_l$ exists if and only if the compatibility condition*

$$f-g\in\{v|_{\partial\Omega};\, v\in\mathcal{H}^{l,q}(\Omega)\}^\circ \tag{5.20}$$

is satisfied; here $1/p+1/q=1$ and the superscript "$\circ$" refers to the annihilator in $L^p(\partial\Omega,\Lambda^l T\mathcal{M})$.

2. *The space of solutions for the homogeneous problem is precisely $\mathcal{H}^{l,p}(\Omega)$. In particular, du and δu are uniquely determined by the boundary data and, further,*

$$\begin{aligned}&\|\mathcal{N}(du)\|_{L^p(\partial\Omega)} + \|\mathcal{N}(\delta u)\|_{L^p(\partial\Omega)}\\ &\qquad \le C\left(\|f\|_{L^p(\partial\Omega,\Lambda^l T\mathcal{M})} + \|g\|_{L^p(\partial\Omega,\Lambda^l T\mathcal{M})}\right)\end{aligned} \tag{5.21}$$

for some positive constant C, independent of f, g.

If the compatibility condition (5.20) is satisfied, we also have:

3. $f\in L^{p,\delta}_{\rm tan}(\partial\Omega,\Lambda^l T\mathcal{M}) \iff \mathcal{N}(\delta du)\in L^p(\partial\Omega)$
$\iff \mathcal{N}(d\delta u)\in L^p(\partial\Omega) \iff g\in L^{p,d}_{\rm nor}(\partial\Omega,\Lambda^l T\mathcal{M})$.

4. $f\in L^{p,0}_{\rm tan}(\partial\Omega,\Lambda^l T\mathcal{M}) \iff \delta du = 0 \iff d\delta u = 0$

$\Longleftrightarrow g \in L^{p,0}_{\text{nor}}(\partial\Omega, \Lambda^l T\mathcal{M})$. *In this case,* $(BVP7)_l$ *becomes*

$$(BVP8)_l \begin{cases} u \in C^1(\Omega, \Lambda^l T\mathcal{M}), \\ \delta du = d\delta u = 0 \text{ in } \Omega, \\ \mathcal{N}(u), \mathcal{N}(du), \mathcal{N}(\delta u) \in L^p(\partial\Omega), \\ \nu \vee (du)|_{\partial\Omega} = f \in L^{p,0}_{\text{tan}}(\partial\Omega, \Lambda^l T\mathcal{M}), \\ \nu \wedge (\delta u)|_{\partial\Omega} = g \in L^{p,0}_{\text{nor}}(\partial\Omega, \Lambda^l T\mathcal{M}). \end{cases}$$

5. $f = 0 \Longleftrightarrow du = 0$. *Granted this,* $(BVP7)_l$ *becomes*

$$(BVP9)_l \begin{cases} u \in C^1(\Omega, \Lambda^l T\mathcal{M}), \\ \Delta u = 0 \text{ in } \Omega, \\ du = 0 \text{ in } \Omega, \\ \mathcal{N}(u),\, \mathcal{N}(\delta u) \in L^p(\partial\Omega), \\ \nu \wedge (\delta u)|_{\partial\Omega} = g \in L^{p,0}_{\text{nor}}(\partial\Omega, \Lambda^l T\mathcal{M}). \end{cases}$$

6. $g = 0 \Longleftrightarrow \delta u = 0$. *In this case,* $(BVP7)_l$ *becomes the Hodge-dual of* $(BVP9)_l$, *i.e.*

$$(BVP10)_l \begin{cases} u \in C^1(\Omega, \Lambda^l T\mathcal{M}), \\ \Delta u = 0 \text{ in } \Omega, \\ \delta u = 0 \text{ in } \Omega, \\ \mathcal{N}(u),\, \mathcal{N}(du) \in L^p(\partial\Omega), \\ \nu \vee (du)|_{\partial\Omega} = f \in L^{p,0}_{\text{tan}}(\partial\Omega, \Lambda^l T\mathcal{M}). \end{cases}$$

The algebraic structure of the Hodge Laplacian (cf. (4.9)) suggests that a natural "conormal derivative" on $\partial\Omega$ is, because of (4.7),

$$\nu \wedge \delta + \nu \vee d.$$

Thus, Theorem 5.2 can be regarded as the natural conormal (or "Neumann") derivative problem for the Hodge Laplacian Δ. In particular, since $\dim \mathcal{H}^{l,p}(\Omega) = \infty$ for $1 \le l \le m-1$, this is an example of a natural Neumann type problem which is not normally solvable (cf. the discussion in the introduction). Let us also point out that, in the smooth context, problems related to $(BVP7)_l$ have been studied by G. F. D. Duff [Du3], P. E. Conner [Co], C. B. Morrey [Mo1], D. C. Spencer [Sp], and G. Schwarz [Sc].

Other natural boundary conditions of interest are discussed in the following three theorems.

THEOREM 5.3. *Let* Ω *be a Lipschitz domain in* $\mathcal{M}$ *and, for each* $0 \le l \le m$, *consider the boundary value problem*

$$(BVP11)_l \begin{cases} u \in C^1(\Omega, \Lambda^l T\mathcal{M}), \\ \Delta u = 0 \text{ in } \Omega, \\ d\delta u = 0 \text{ in } \Omega, \\ \mathcal{N}(u), \mathcal{N}(du), \mathcal{N}(\delta u) \in L^p(\partial\Omega), \\ \nu \vee (\delta u)|_{\partial\Omega} = f \in L^p_{\text{tan}}(\partial\Omega, \Lambda^{l-2} T\mathcal{M}), \\ \nu \wedge (du)|_{\partial\Omega} = g \in L^p_{\text{nor}}(\partial\Omega, \Lambda^{l+2} T\mathcal{M}). \end{cases}$$

There exists $\varepsilon = \varepsilon(\Omega) > 0$ so that, if $2-\varepsilon < p < 2+\varepsilon$, then a solution to $(BVP11)_l$ exists only if the compatibility conditions

$$f \in \delta_\partial L^{p,\delta}_{\rm tan}(\partial\Omega, \Lambda^{l-1}T\mathcal{M}), \quad g \in d_\partial L^{p,d}_{\rm nor}(\partial\Omega, \Lambda^{l+1}T\mathcal{M}) \tag{5.22}$$

are satisfied. Moreover, the space of null solutions for $(BVP11)_l$ is infinite dimensional.

THEOREM 5.4. *Let Ω be a Lipschitz domain in $\mathcal{M}$ and, for each $0 \le l \le m$, consider the boundary value problem*

$$(BVP12)_l \begin{cases} u \in C^1(\Omega, \Lambda^l T\mathcal{M}), \\ \Delta u = 0 \ in \ \Omega, \\ d\delta u = 0 \ in \ \Omega, \\ \mathcal{N}(u), \mathcal{N}(du) \in L^p(\partial\Omega), \\ \nu \vee u|_{\partial\Omega} = f \in L^p_{\rm tan}(\partial\Omega, \Lambda^{l-1}T\mathcal{M}), \\ \nu \wedge (du)|_{\partial\Omega} = g \in L^p_{\rm nor}(\partial\Omega, \Lambda^{l+2}T\mathcal{M}). \end{cases}$$

There exists $\varepsilon = \varepsilon(\Omega) > 0$ so that, if $2-\varepsilon < p < 2+\varepsilon$, then $(BVP12)_l$ admits a solution if and only if

$$g \in d_\partial L^{p,d}_{\rm nor}(\partial\Omega, \Lambda^{l+1}T\mathcal{M}). \tag{5.23}$$

Furthermore, the space of null solutions for $(BVP12)_l$ is finite dimensional.

Similar results are valid for the Hodge dual of $(BVP12)_l$, i.e.

$$(BVP13)_l \begin{cases} u \in C^1(\Omega, \Lambda^l T\mathcal{M}), \\ \Delta u = 0 \ in \ \Omega, \\ d\delta u = 0 \ in \ \Omega, \\ \mathcal{N}(u), \mathcal{N}(\delta u) \in L^p(\partial\Omega), \\ \nu \wedge u|_{\partial\Omega} = f \in L^p_{\rm nor}(\partial\Omega, \Lambda^{l+1}T\mathcal{M}), \\ \nu \vee (\delta u)|_{\partial\Omega} = g \in L^p_{\rm tan}(\partial\Omega, \Lambda^{l-2}T\mathcal{M}). \end{cases}$$

THEOREM 5.5. *Let Ω be a Lipschitz domain in $\mathcal{M}$ and, for each $0 \le l \le m$, consider the boundary value problem*

$$(BVP14)_l \begin{cases} u \in C^1(\Omega, \Lambda^l T\mathcal{M}), \\ \Delta u = 0 \ in \ \Omega, \\ \mathcal{N}(u), \mathcal{N}(du), \mathcal{N}(\delta u) \in L^p(\partial\Omega), \\ \nu \wedge u|_{\partial\Omega} = f \in L^{p,d}_{\rm nor}(\partial\Omega, \Lambda^{l+1}T\mathcal{M}), \\ \nu \vee (du)|_{\partial\Omega} = g \in L^p_{\rm tan}(\partial\Omega, \Lambda^l T\mathcal{M}). \end{cases}$$

There exists $\varepsilon = \varepsilon(\Omega) > 0$ so that, if $2-\varepsilon < p < 2+\varepsilon$, then a solution to $(BVP14)_l$ exists only if the compatibility condition

$$\int_{\partial\Omega} \langle f, d\omega \rangle \, d\sigma = \int_{\partial\Omega} \langle g, \omega \rangle \, d\sigma \tag{5.24}$$

holds for each form ω *satisfying*

$$
\begin{cases}
\omega \in C^1(\Omega, \Lambda^l T\mathcal{M}),\\
\Delta\omega = 0 \ in\ \Omega,\\
\delta\omega = 0 \ in\ \Omega,\\
\mathcal{N}(\omega), \mathcal{N}(d\omega) \in L^q(\partial\Omega),\\
\nu \vee \omega|_{\partial\Omega} = 0 \ on\ \partial\Omega,
\end{cases}
\tag{5.25}
$$

where $1/p + 1/q = 1$. *Moreover,*

$$
du = 0 \Leftrightarrow f = 0 \ and\ g = 0.
\tag{5.26}
$$

In particular, the space of null solutions for $(BVP14)_l$ *is characterized by*

$$
\begin{cases}
u \in C^1(\Omega, \Lambda^l T\mathcal{M}),\\
\Delta u = 0 \ in\ \Omega,\\
du = 0 \ in\ \Omega,\\
\mathcal{N}(u),\ \mathcal{N}(\delta u) \in L^p(\partial\Omega),\\
\nu \wedge u|_{\partial\Omega} = 0 \ on\ \partial\Omega.
\end{cases}
\tag{5.27}
$$

Finally, an analogous set of results is valid for the Hodge dual problem, i.e. for

$$
(BVP15)_l
\begin{cases}
w \in C^1(\Omega, \Lambda^l T\mathcal{M}),\\
\Delta w = 0 \ in\ \Omega,\\
\mathcal{N}(w), \mathcal{N}(dw), \mathcal{N}(\delta w) \in L^p(\partial\Omega),\\
\nu \vee w|_{\partial\Omega} = f \in L^{p,\delta}_{\rm tan}(\partial\Omega, \Lambda^{l-1} T\mathcal{M}),\\
\nu \wedge (\delta w)|_{\partial\Omega} = g \in L^p_{\rm nor}(\partial\Omega, \Lambda^l T\mathcal{M}).
\end{cases}
$$

Essentially, the rest of the monograph (with the exception of §14) is devoted to presenting the proofs of the theorems stated in this chapter (final arguments are given in §§ 12 – 13). Our strategy is to work with boundary layer operators, well adapted for the problems at hand, and the task of developing these tools systematically is taken up in the next chapter.

CHAPTER 6

Layer Potential Operators on Lipschitz Domains

Let $V \geq 0$ be a C^1 scalar-valued function not identically zero. Recall that under these conditions, the operator $L := \Delta - V$, acting on suitable spaces of l-forms has an inverse, $(\Delta - V)^{-1}$, whose Schwartz kernel, $\Gamma_l(x, y)$, is a symmetric double form of bidegree (l, l). In local coordinates $\Gamma_l(x, y)$ satisfies

$$
(6.1) \qquad \begin{aligned} \Gamma_l(x,y) = C_m \frac{1}{\sqrt{g(y)}} &\left(\sum_{j,k} g_{jk}(y)(x_j - y_j)(x_k - y_k) \right)^{-(m-2)/2} \\ &\times \sum_{|I|=l} \sum_{|J|=l} \det\left((g_{ij}(y))_{i \in I, j \in J} \right) dx^I \otimes dy^J \\ &+ \text{a less singular term.} \end{aligned}
$$

This follows from (2.21), (3.17) and (4.1); note that the remainder in the above expansion satisfies (3.18). It is also useful to observe that, since Δ and $*$ commute, Lemma 4.1 gives

$$
(6.2) \qquad *_x *_y \Gamma_l(x, y) = \Gamma_{m-l}(x, y),
$$

where the subscripts indicate the variables in which the corresponding (Hodge star-) operators are acting.

Further, from $\delta(\Delta - V) = (\Delta - V)\delta - dV \vee \cdot$, it follows that

$$
\delta(\Delta - V)^{-1} = (\Delta - V)^{-1}\delta + (\Delta - V)^{-1}(dV\vee)(\Delta - V)^{-1}.
$$

At the level of Schwartz kernels, this identity amounts to

$$
(6.3) \qquad \delta_x(\Gamma_{l+1}(x, y)) = d_y(\Gamma_l(x, y)) + R_l(x, y)
$$

where the residual kernel $R_l(x, y)$ is a double form of bidegree $(l, l+1)$ which, in local coordinates, is $\mathcal{O}(|x - y|^{-(m-4)})$ as $|x - y| \to 0$. Note that $L_x R_l(x, y) = dV(x) \vee \Gamma_{l+1}(x, y)$, and

$$
R_l \in C^1_{\text{loc}}\big((\mathcal{M} \times \mathcal{M} \setminus \text{diag}) \cup \{(x, y) : x \notin \text{supp } dV\}\big)
$$

We shall occasionally also use R_l to denote the boundary integral operator with kernel $R_l(x, y)$, that is

$$
(6.4) \qquad R_l f(x) := \int_{\partial\Omega} \langle R_l(x, y), f(y) \rangle \, d\sigma(y), \quad f \in L^p(\partial\Omega, \Lambda^{l+1}T\mathcal{M}).
$$

Analogously,

$$
(6.5) \qquad d_x(\Gamma_l(x, y)) = \delta_y(\Gamma_{l+1}(x, y)) + Q_l(x, y)
$$

where $Q_l(x,y)$ is a double form of bidegree $(l+1,l)$ which exhibits a similar behavior; in fact $Q_l(x,y) = -R_l(y,x)$. Once again, we shall utilize Q_l for the boundary integral operator with kernel $Q_l(x,y)$ also.

Next, let Ω be a Lipschitz domain in $\mathcal{M}$ and denote by $\mathcal{S}_l$ the single layer potential operator on $\partial\Omega$ with kernel $\Gamma_l(x,y)$, i.e.,

$$\mathcal{S}_l f(x) := \int_{\partial\Omega} \langle \Gamma_l(x,y), f(y)\rangle\, d\sigma(y), \quad x \in \mathcal{M}\setminus\partial\Omega, \tag{6.6}$$

where $f \in L^p(\partial\Omega, \Lambda^l T\mathcal{M})$. Note that $(\Delta - V)\mathcal{S}_l f = 0$ in $\mathcal{M}\setminus\partial\Omega$. Also, set $S_l f := \mathcal{S}_l f|_{\partial\Omega}$. Theorem 2.9 implies that for $1 < p < \infty$ we have

$$\|\mathcal{N}(S_l f)\|_{L^p(\partial\Omega)},\ \|\mathcal{N}(d\mathcal{S}_l f)\|_{L^p(\partial\Omega)},\ \|\mathcal{N}(\delta\mathcal{S}_l f)\|_{L^p(\partial\Omega)} \leq C\|f\|_{L^p(\partial\Omega,\Lambda^l T\mathcal{M})} \tag{6.7}$$

uniformly for $f \in L^p(\partial\Omega, \Lambda^l T\mathcal{M})$. Also, on account of Theorem 2.9 and (4.7),

$$\begin{aligned} d\mathcal{S}_l f|_{\partial\Omega_\pm}(x) &= \mp\tfrac{1}{2}\nu(x)\wedge f(x) + \text{p.v.}\int_{\partial\Omega}\langle d_x\Gamma_l(x,y), f(y)\rangle\, d\sigma(y),\\ \delta\mathcal{S}_l f|_{\partial\Omega_\pm}(x) &= \pm\tfrac{1}{2}\nu(x)\vee f(x) + \text{p.v.}\int_{\partial\Omega}\langle \delta_x\Gamma_l(x,y), f(y)\rangle\, d\sigma(y), \end{aligned} \tag{6.8}$$

for a.e. $x \in \partial\Omega$. In the sequel, we shall denote by dS_l and δS_l the boundary principal-value integral operators appearing above, that is, the p.v. operators with kernels $d_x\Gamma_l(x,y)$ and $\delta_x\Gamma_l(x,y)$, respectively. In particular, $d\mathcal{S}_l|_{\partial\Omega_\pm} = \mp\frac{1}{2}\nu\wedge\cdot + dS_l$ and $\delta\mathcal{S}_l|_{\partial\Omega_\pm} = \pm\frac{1}{2}\nu\vee\cdot + \delta S_l$.

We also introduce the closely related operators

$$\begin{aligned} M_l f(x) &:= \nu(x)\vee \text{p.v.}\int_{\partial\Omega}\langle d_x\Gamma_l(x,y), f(y)\rangle d\sigma(y), \quad x\in\partial\Omega,\\ N_l f(x) &:= \nu(x)\wedge \text{p.v.}\int_{\partial\Omega}\langle \delta_x\Gamma_l(x,y), f(y)\rangle d\sigma(y), \quad x\in\partial\Omega, \end{aligned} \tag{6.9}$$

and

$$\begin{aligned} \tilde{M}_l f(x) &:= \nu(x)\vee \text{p.v.}\int_{\partial\Omega}\langle \delta_y\Gamma_{l+1}(x,y), f(y)\rangle\, d\sigma(y), \quad x\in\partial\Omega,\\ \tilde{N}_l g(x) &:= \nu(x)\wedge \text{p.v.}\int_{\partial\Omega}\langle d_y\Gamma_{l-1}(x,y), g(y)\rangle\, d\sigma(y), \quad x\in\partial\Omega. \end{aligned} \tag{6.10}$$

In passing, let us note that $V \equiv$ constant forces $\tilde{M}_l = M_l$ and $\tilde{N}_l = N_l$. For further reference, we note here that the jump-relations (6.8) give

$$\nu\vee d\mathcal{S}_l f|_{\partial\Omega_\pm} = \mp\tfrac{1}{2}f_{\tan} + M_l f, \quad \nu\wedge\delta\mathcal{S}_l f|_{\partial\Omega_\pm} = \pm\tfrac{1}{2}f_{\text{nor}} + N_l f. \tag{6.11}$$

Other basic properties of these operators are summarized in the next proposition.

PROPOSITION 6.1. *For each $l = 0,1,\dots,m$, $1 < p < \infty$, the operators M_l, N_l are bounded mappings from $L^p(\partial\Omega,\Lambda^l T\mathcal{M})$ into $L^p_{\tan}(\partial\Omega,\Lambda^l T\mathcal{M})$ and from $L^p(\partial\Omega,\Lambda^l T\mathcal{M})$ into $L^p_{\text{nor}}(\partial\Omega,\Lambda^l T\mathcal{M})$, respectively. Also,*

$$M_{m-l}* = -*N_l \quad \text{and} \quad *M_l = -N_{m-l}* \quad \text{on } l\text{-forms}. \tag{6.12}$$

Furthermore, the adjoint of M_l acting on $L^p_{\tan}(\partial\Omega,\Lambda^l T\mathcal{M})$ is the operator M_l^t acting on $L^q_{\tan}(\partial\Omega,\Lambda^l T\mathcal{M})$, with $1/p+1/q=1$, given by

$$M_l^t = \nu\vee\tilde{N}_{l+1}(\nu\wedge\cdot), \tag{6.13}$$

while the adjoint of $\tilde{M}_l$ acting on $L^p_{\tan}(\partial\Omega, \Lambda^l T\mathcal{M})$ is the operator $\tilde{M}_l^t$ acting on the space $L^q_{\tan}(\partial\Omega, \Lambda^l T\mathcal{M})$ by

$$\tilde{M}_l^t = \nu \vee N_{l+1}(\nu \wedge \cdot). \tag{6.14}$$

Finally, if supp$(dV) \cap \partial\Omega = \varnothing$ then M_l, N_l are well-defined, bounded mappings of $L^{p,\delta}_{\tan}(\partial\Omega, \Lambda^l T\mathcal{M})$ and $L^{p,d}_{\text{nor}}(\partial\Omega, \Lambda^l T\mathcal{M})$, respectively, into themselves.

PROOF. The first part of the proposition is a direct consequence of Theorem 2.9, while (6.12) follows from (6.2) and the various intertwining properties of the Hodge star-isomorphism; cf. also Lemma 4.1. Going further, let $\Omega_j \nearrow \Omega$ be a nested sequence of Lipschitz domains exhausting Ω in a convenient way and denote by ν_j, $d\sigma_j$, respectively, the outward unit conormal and the surface measure for $\partial\Omega_j$. If $\phi, \psi \in C^0(\mathcal{M}, \Lambda^l T\mathcal{M})$ are arbitrary we have

$$\begin{aligned}
&\int_{\partial\Omega} \langle (-\tfrac{1}{2}I + M_l)(\phi_{\tan}), \psi_{\tan}\rangle \, d\sigma \\
&= \lim_{j\to\infty} \int_{\partial\Omega_j} \langle d\mathcal{S}_l(\phi_{\tan}), \nu_j \wedge \psi\rangle \, d\sigma_j \\
&= \lim_{j\to\infty} \int_{\partial\Omega}\int_{\partial\Omega_j} \langle\langle d_x\Gamma_l(x,y), \nu_j(x) \wedge \psi(x)\rangle, \phi_{\tan}(y)\rangle \, d\sigma_j(x) d\sigma(y) \\
&= \lim_{j\to\infty} \int_{\partial\Omega}\int_{\partial\Omega_j} \langle\langle \delta_y\Gamma_{l+1}(x,y), \nu_j(x) \wedge \psi(x)\rangle, \phi_{\tan}(y)\rangle \, d\sigma_j(x) d\sigma(y) \\
&\quad + \lim_{j\to\infty} \int_{\partial\Omega}\int_{\partial\Omega_j} \langle\langle Q_l(x,y), \nu_j(x) \wedge \psi(x)\rangle, \phi_{\tan}(y)\rangle \, d\sigma_j(x) d\sigma(y) \\
&= \int_{\partial\Omega} \langle \nu \vee (\nu \wedge \delta\mathcal{S}_{l+1}(\nu \wedge \psi_{\tan}))\,|_{\partial\Omega_-}, \phi_{\tan}\rangle \, d\sigma \\
&\quad - \int_{\partial\Omega} \langle \nu \vee (\nu \wedge R_l(\nu \wedge \psi_{\tan}))\,, \phi_{\tan}\rangle \, d\sigma.
\end{aligned}$$

Then (6.13) follows from (6.11), (6.3) and a density argument; note that we use the identification $\left(L^p_{\tan}(\partial\Omega, \Lambda^l T\mathcal{M})\right)^* = L^q_{\tan}(\partial\Omega, \Lambda^l T\mathcal{M})$ for $1 < p, q < \infty$, $1/p + 1/q = 1$. A similar reasoning applies to (6.14).

Turning attention to the last part of the theorem, we shall need to derive certain identities of independent interest. First, so we claim, if $u \in C^1(\Omega, \Lambda^l T\mathcal{M})$ is such that $\mathcal{N}(u), \mathcal{N}(\delta u) \in L^p(\partial\Omega)$ for some $p \in [1,\infty)$ and $u, \delta u$ have non-tangential boundary traces at almost any point on $\partial\Omega$, then $\nu \vee u|_{\partial\Omega} \in L^p_{\tan}(\partial\Omega, \Lambda^{l-1} T\mathcal{M})$ and

$$\delta_\partial(\nu \vee u|_{\partial\Omega}) = -\nu \vee (\delta u)|_{\partial\Omega}. \tag{6.15}$$

Similarly, if u and du have non-tangential boundary traces at almost any point on $\partial\Omega$ and $\mathcal{N}(u), \mathcal{N}(du) \in L^p(\partial\Omega)$ for some $1 \leq p < \infty$, then $\nu \wedge u|_{\partial\Omega}$ belongs to $L^p_{\text{nor}}(\partial\Omega, \Lambda^{l+1} T\mathcal{M})$ and in fact

$$d_\partial(\nu \wedge u|_{\partial\Omega}) = -\nu \wedge (du)|_{\partial\Omega}. \tag{6.16}$$

To prove, for instance, the identity (6.15), let $\psi \in C^1(\mathcal{M}, \Lambda^{l-2}T\mathcal{M})$ be arbitrary and assume that $u \in C^1(\overline{\Omega}, \Lambda^l T\mathcal{M})$. Successive integrations by parts give

$$\begin{aligned}\int_{\partial\Omega} \langle d\psi, \nu \vee u\rangle \, d\sigma &= \iint_{\Omega} \langle dd\psi, u\rangle \, d\mathrm{Vol} - \iint_{\Omega} \langle d\psi, \delta u\rangle \, d\mathrm{Vol} \\ &= -\iint_{\Omega} \langle \psi, \delta\delta u\rangle \, d\mathrm{Vol} - \int_{\partial\Omega} \langle \nu \wedge \psi, \delta u\rangle \, d\sigma \\ &= -\int_{\partial\Omega} \langle \psi, \nu \vee \delta u\rangle \, d\sigma.\end{aligned}$$

In fact, upon closer inspection of the proof, we see that the equality $\int_{\partial\Omega}\langle d\psi, \nu \vee u\rangle \, d\sigma = -\int_{\partial\Omega}\langle \psi, \nu \vee \delta u\rangle \, d\sigma$ continues to be valid under the weaker condition on u stated above. Now, the formula (6.15) follows from the definition of δ_∂. A similar proof works for (6.16).

The second set of identities alluded to earlier are

$$\begin{aligned}&\delta \mathcal{S}_l f = \mathcal{S}_{l-1}(\delta_\partial f) + R_{l-1} f, \quad \forall f \in L^{p,\delta}_{\tan}(\partial\Omega, \Lambda^l T\mathcal{M}), \\ &d\mathcal{S}_l g = \mathcal{S}_{l+1}(d_\partial g) + Q_l g, \quad \forall g \in L^{p,d}_{\mathrm{nor}}(\partial\Omega, \Lambda^l T\mathcal{M}).\end{aligned} \tag{6.17}$$

To see the first equality, let $f \in L^{p,\delta}_{\tan}(\partial\Omega, \Lambda^l T\mathcal{M})$ be arbitrary. On account of (6.3) and (5.1), we may write

$$\begin{aligned}\delta \mathcal{S}_l f(x) &= \int_{\partial\Omega} \langle \delta_x(\Gamma_l(x,y)), f(y)\rangle \, d\sigma(y) \\ &= \int_{\partial\Omega} \langle d_y(\Gamma_{l-1}(x,y)), f(y)\rangle \, d\sigma(y) + \int_{\partial\Omega} \langle R_{l-1}(x,y), f(y)\rangle \, d\sigma(y) \\ &= \int_{\partial\Omega} \langle \Gamma_{l-1}(x,y), (\delta_\partial f)(y)\rangle \, d\sigma(y) + \int_{\partial\Omega} \langle R_{l-1}(x,y), f(y)\rangle \, d\sigma(y) \\ &= \mathcal{S}_{l-1}(\delta_\partial f)(x) + R_{l-1} f(x),\end{aligned} \tag{6.18}$$

i.e. the first identity in (6.17). The second identity in (6.17) follows by a similar argument.

We are now in a position to tackle the last part in the statement of Proposition 6.1. This will, in fact, be a direct consequence of the fact that M_l, N_l "essentially" commute with the boundary derivatives operators δ_∂ and d_∂, respectively. More concretely, we shall prove that for any $f \in L^{p,\delta}_{\tan}(\partial\Omega, \Lambda^l T\mathcal{M})$ we have

$$\begin{aligned}\delta_\partial M_l f(x) = M_{l-1}(\delta_\partial f)(x) &+ \nu(x) \vee V(x) \mathcal{S}_l f(x) \\ &+ \nu(x) \vee \int_{\partial\Omega} \langle d_x R_{l-1}(x,y), f(y)\rangle \, d\sigma(y),\end{aligned} \tag{6.19}$$

while for any $g \in L^{p,d}_{\mathrm{nor}}(\partial\Omega, \Lambda^l T\mathcal{M})$ we have

$$\begin{aligned}d_\partial N_l g(x) = N_{l+1}(d_\partial g)(x) &+ \nu(x) \wedge V(x) \mathcal{S}_l g(x) \\ &+ \nu(x) \wedge \int_{\partial\Omega} \langle \delta_x Q_l(x,y), g(y)\rangle \, d\sigma(y).\end{aligned} \tag{6.20}$$

Indeed, set $u := d\mathcal{S}_l f \in C^1(\Omega, \Lambda^{l+1} T\mathcal{M})$. Clearly, $\mathcal{N}(u) \in L^p(\partial\Omega)$ and since by the commutation formula (6.17)

$$\delta u = \delta d \mathcal{S}_l f = -V\mathcal{S}_l f - d\delta \mathcal{S}_l f = -V\mathcal{S}_l f - d\mathcal{S}_l(\delta_\partial f) - dR_{l-1} f \tag{6.21}$$

(hereafter, for each l, dR_l will denote the boundary principal-value operator with kernel $d_x R_l(x,y)$, etc.) we see that $\mathcal{N}(\delta u) \in L^p(\partial\Omega)$ also. Also, from (6.11) we have that $\nu \vee u|_{\partial\Omega} = \left(-\frac{1}{2}I + M_l\right) f$ and, hence,

$$\begin{aligned}
\delta_\partial M_l f &= \tfrac{1}{2}\delta_\partial f + \delta_\partial \left(\nu \vee u|_{\partial\Omega}\right) \\
&= \tfrac{1}{2}\delta_\partial f - \nu \vee (\delta u)|_{\partial\Omega} && \text{(by (6.15))} \\
&= \tfrac{1}{2}\delta_\partial f + \nu \vee dS_l(\delta_\partial f)|_{\partial\Omega} + \nu \vee VS_l f + \nu \vee dR_{l-1} f && \text{(by (6.21))} \\
&= \tfrac{1}{2}\delta_\partial f - \tfrac{1}{2}\delta_\partial f + M_{l-1}(\delta_\partial f) + \nu \vee VS_l f + \nu \vee dR_{l-1} f && \text{(by (6.11))} \\
&= M_{l-1}(\delta_\partial f) + \nu \vee VS_l f + \nu \vee dR_{l-1} f.
\end{aligned}$$

This proves (6.19). Since (6.20) is proved similarly the proof of Proposition 6.1 is completed. ■

Next we shall turn attention to studying finer spectral properties of the operators M_l, N_l. In this respect, our main result is contained in the next theorem.

THEOREM 6.2. *Let $\mathcal{M}$, Ω be as before and consider a positive potential $V \in L^\infty(\mathcal{M})$ which is > 0 on some open subset of $\mathcal{M}$. Then there exists $\varepsilon = \varepsilon(\Omega, V) > 0$ so that for every $0 \le l \le m$ the operators*

$$\pm\tfrac{1}{2}I + M_l : L^p_{\tan}(\partial\Omega, \Lambda^l T\mathcal{M}) \to L^p_{\tan}(\partial\Omega, \Lambda^l T\mathcal{M})$$

and

$$\pm\tfrac{1}{2}I + N_l : L^p_{\mathrm{nor}}(\partial\Omega, \Lambda^l T\mathcal{M}) \to L^p_{\mathrm{nor}}(\partial\Omega, \Lambda^l T\mathcal{M})$$

are Fredholm with index zero whenever $2 - \varepsilon < p < 2 + \varepsilon$.

The proof of this result is rather lengthy so it is worthwhile to sketch briefly some of the ideas involved. In the first stage we shall prove a weaker version of Theorem 6.2 to the effect that $\pm\frac{1}{2}I + M_l$, $\pm\frac{1}{2}I + N_l$ are Fredholm with index zero when acting on spaces of *regular* forms, i.e. $L^{p,\delta}_{\tan}(\partial\Omega, \Lambda^l T\mathcal{M})$ and $L^{p,d}_{\mathrm{nor}}(\partial\Omega, \Lambda^l T\mathcal{M})$, respectively. This is accomplished by means of certain new boundary energy estimates which, in turn, are derived from some integral identities (generically referred to in the field as Rellich identities) which we devise in §7. In fact, as we shall see early in § 8, if $l = 0$ or $l = 1$, these identities yield the theorem. The case $l = 0$, when M_l can be identified with the more familiar adjoint of the double layer potential for the Laplacian, corresponds to the scalar-valued theory developed in the Euclidean setting by G. Verchota [Ve1], B. Dahlberg and C. Kenig [DK]; cf. also the work of M. Mitrea and M. Taylor [MT1] in the context of Riemannian manifolds. Although, strictly speaking, $L^p_{\tan}(\partial\Omega, \Lambda^0 T\mathcal{M}) \equiv L^p(\partial\Omega)$ is not a space of "regular" forms in the usual sense, the smallness of the degree is an appropriate substitute; indeed, it is because of simple degree considerations that the most difficult terms to control in the key estimate drop out in this case. A similar phenomenon occurs for $l = 1$ which is a setting closely related to problems in electromagnetism. This remarkable "accident" was first employed (in the flat, Euclidean setting) by D. Mitrea, M. Mitrea and J. Pipher [MMP]. Also, note that if $l = m$ then M_l vanishes identically. However, away from these extreme values of l new ideas are required.

The next step, taken in § 10, is to use a localization argument to reduce the proof of Theorem 6.2 to the case when $\mathcal{M}$ is a *homology sphere*, so that all singular homology groups $H^q_{\mathrm{sing}}(\mathcal{M}; \mathbb{R})$ vanish (for $0 < q < m$). The important thing here is that the main singularity in the Schwartz kernel of $(\Delta - V)^{-1}$ depends exclusively on the metric tensor and neither on V nor on the topology. The main feature of working on homology spheres is that it enables us to dispense with V altogether (by taking $V = 0$); in particular, $L^{p,0}_{\tan}(\partial\Omega, \Lambda^l T\mathcal{M})$ (see (5.5)) becomes an invariant subspace of M_l, and $\frac{1}{2}I + M_l$ acting on $L^{p,0}_{\tan}(\partial\Omega, \Lambda^l T\mathcal{M})$ is Fredholm with index zero.

Retaining the special topological hypothesis from the previous step we then proceed to establish the missing link, i.e., that $\frac{1}{2}I + M_l$ acting on the *quotient* space

$$\frac{L^p_{\tan}(\partial\Omega, \Lambda^l T\mathcal{M})}{L^{p,0}_{\tan}(\partial\Omega, \Lambda^l T\mathcal{M})}$$

is Fredholm with index zero. This is done by certain rather technical duality arguments that heavily rely on tools from algebraic topology. Finally, with this in hand, we are able then to restore the original context, i.e., working in the space $L^p_{\tan}(\partial\Omega, \Lambda^l T\mathcal{M})$ "at large" via abstract functional analytic methods (this somewhat vague outline is made precise in § 10).

Accomplishing this program takes up the bulk of the next four consecutive chapters and final arguments in the proof of Theorem 6.2 are presented in § 10.

CHAPTER 7

Rellich Type Estimates for Differential Forms

The results in this chapter are valid for any smoothly differentiable compact manifold $\mathcal{M}$ of real dimension m, which we consider equipped with a Lipschitzian Riemannian metric, and any fixed potential $V \in L^\infty(\mathcal{M})$.

PROPOSITION 7.1. *Let Ω be an arbitrary Lipschitz domain in $\mathcal{M}$ and assume that θ is an arbitrary C^1-smooth one form on $\mathcal{M}$. Then, for $0 \le l \le m$, and for any l-form $u \in C^1(\bar{\Omega}, \Lambda^l T\mathcal{M})$ that verifies $(\Delta - V)u = 0$ in Ω, we have*

$$\begin{aligned}\int_{\partial\Omega} \langle \theta \vee du, \nu \vee du \rangle \, d\sigma + \int_{\partial\Omega} \langle \theta \vee \delta u, \nu \vee \delta u \rangle \, d\sigma - \int_{\partial\Omega} \langle \nu \wedge \delta u, \theta \vee du \rangle \, d\sigma \\ = \int_{\partial\Omega} \tfrac{1}{2}|du|^2 \langle \theta, \nu \rangle \, d\sigma + \int_{\partial\Omega} \tfrac{1}{2}|\delta u|^2 \langle \theta, \nu \rangle \, d\sigma + R(\theta, u),\end{aligned} \tag{7.1}$$

where the "residue" $R(\theta, u)$ satisfies

$$|R(\theta, u)| \le C(\|V\|_{L^\infty} + \|\theta\|_{L^\infty} + \|\nabla\theta\|_{L^\infty}) \iint_\Omega (|u|^2 + |du|^2 + |\delta u|^2) \, d\,Vol \tag{7.2}$$

uniformly in u, V and θ.

PROOF. The departure point is to note that for any C^1 differential form ω there holds

$$\begin{aligned}\int_{\partial\Omega} \tfrac{1}{2}|\omega|^2 \langle \theta, \nu \rangle \, d\sigma - \int_{\partial\Omega} \langle \theta \vee \omega, \nu \vee \omega \rangle \, d\sigma \\ = \iint_\Omega \langle \delta\omega, \theta \vee \omega \rangle \, d\mathrm{Vol} + \iint_\Omega \langle d\omega, \theta \wedge \omega \rangle \, d\mathrm{Vol} + \tilde{R}(\theta, \omega),\end{aligned} \tag{7.3}$$

where

$$|\tilde{R}(\theta, \omega)| \le C \iint_\Omega |\nabla\theta| \cdot |\omega|^2 \, d\mathrm{Vol}. \tag{7.4}$$

To see this, we use Lemma 2.1 and repeated integration by parts to transform the left side of (7.3) into

$$\begin{aligned}
\tfrac{1}{2}\int_{\partial\Omega}\langle\theta\wedge(\nu\vee\omega)+\nu\vee(\theta\wedge\omega),\,\omega\rangle\,d\sigma-\int_{\partial\Omega}\langle\theta\vee\omega,\nu\vee\omega\rangle\,d\sigma \\
=\tfrac{1}{2}\int_{\partial\Omega}\langle\nu\wedge\omega,\theta\wedge\omega\rangle\,d\sigma-\tfrac{1}{2}\int_{\partial\Omega}\langle\theta\vee\omega,\nu\vee\omega\rangle\,d\sigma \\
=\tfrac{1}{2}\iint_{\Omega}\langle d\omega,\theta\wedge\omega\rangle\,d\mathrm{Vol}-\tfrac{1}{2}\iint_{\Omega}\langle\omega,\delta(\theta\wedge\omega)\rangle\,d\mathrm{Vol} \\
+\tfrac{1}{2}\iint_{\Omega}\langle\delta\omega,\theta\vee\omega\rangle\,d\mathrm{Vol}-\tfrac{1}{2}\iint_{\Omega}\langle\omega,d(\theta\vee\omega)\rangle\,d\mathrm{Vol} \\
=\iint_{\Omega}\langle\delta\omega,\theta\vee\omega\rangle\,d\mathrm{Vol}+\iint_{\Omega}\langle d\omega,\theta\wedge\omega\rangle\,d\mathrm{Vol}+\tilde{R}(\theta,\omega)
\end{aligned}$$

where the residual term $\tilde{R}(\theta,\omega)$ involves the L^2-pairing (over Ω) of ω with

$$d(\theta\vee\omega)+\theta\vee d\omega+\theta\wedge\delta\omega+\delta(\theta\wedge\omega). \tag{7.5}$$

Now the key element is that, due to internal cancellations, the mapping taking ω into (7.5) is a *zero-order* differential operator so that the expression in (7.5) is $\mathcal{O}(|\omega|)$. Indeed, its *principal symbol* at $\xi\in T_x^*\mathcal{M}$ is precisely

$$i\xi\wedge(\theta\vee\cdot)+i\theta\vee(\xi\wedge\cdot)-i\theta\wedge(\xi\vee\cdot)-i\xi\vee(\theta\wedge\cdot)=i\langle\xi,\theta\rangle_x-i\langle\theta,\xi\rangle_x=0;$$

cf. also (3.15). Therefore, $\tilde{R}(\theta,\omega)=\iint_\Omega\mathcal{O}(|\omega||\nabla\theta|)\,d\mathrm{Vol}$, and this completes the proof of (7.3)–(7.4).

Next, assume $0\le l\le m$ and $u\in C^1(\bar{\Omega},\Lambda^lT\mathcal{M})$ with $(\Delta-V)u=0$ in Ω. The proof of the identity (7.1) will consist in applying (7.3) *twice* to suitable differential forms and then combining the results.

First, we utilize (7.3) for the $(l+1)$-form $\omega:=du$ in Ω and obtain:

$$\int_{\partial\Omega}\tfrac{1}{2}|du|^2\langle\theta,\nu\rangle\,d\sigma-\int_{\partial\Omega}\langle\theta\vee du,\nu\vee du\rangle\,d\sigma=\iint_{\Omega}\langle\delta du,\theta\vee du\rangle\,d\mathrm{Vol}+R(\theta,u) \tag{7.6}$$

where $R(\theta,u)$ can be estimated as in (7.2). Now, using the fact that $\delta du=-d\delta u-Vu$ and integrating by parts, we may recast the right side of the identity (7.6) in the form

$$\begin{aligned}
\iint_{\Omega}\langle\delta du,\theta\vee du\rangle\,d\mathrm{Vol}=-\iint_{\Omega}\langle\delta u,\delta(\theta\vee du)\rangle\,d\mathrm{Vol}-\iint_{\Omega}\langle Vu,\theta\vee du\rangle\,d\mathrm{Vol} \\
-\int_{\partial\Omega}\langle\nu\wedge\delta u,\theta\vee du\rangle\,d\sigma.
\end{aligned} \tag{7.7}$$

In the right side of (7.7), the boundary integral is part of the final identity while the second solid integral can be absorbed in $R(\theta,u)$; we shall further transform the first solid one. To this end, a straightforward calculation shows that

$$\begin{aligned}
\delta(\theta\vee du)&=(-1)^{(m+1)(l+1)}*(d\theta\wedge *du)-\theta\vee\delta du \\
&=(-1)^{(m+1)(l+1)}*(d\theta\wedge *du)+\theta\vee d\delta u+\theta\vee Vu
\end{aligned}$$

and we shall find it convenient to write

$$
\begin{aligned}
-\iint_{\Omega}\langle\delta u,\delta(\theta\vee du)\rangle\, d\mathrm{Vol} &= \iint_{\Omega}\langle *\delta u, d\theta\wedge *du\rangle\, d\mathrm{Vol} - \iint_{\Omega}\langle\theta\wedge\delta u, d\delta u\rangle\, d\mathrm{Vol} \\
&\quad - \iint_{\Omega}\langle\delta u,\theta\vee Vu\rangle\, d\mathrm{Vol}.
\end{aligned} \tag{7.8}
$$

If we now once again apply the identity (7.3), this time to the $(l-1)$-form $\omega := \delta u$, it follows immediately that

$$
\iint_{\Omega}\langle\theta\wedge\delta u, d\delta u\rangle\, d\mathrm{Vol} = \int_{\partial\Omega}\tfrac{1}{2}|\delta u|^2\langle\theta,\nu\rangle\, d\sigma - \int_{\partial\Omega}\langle\theta\vee\delta u,\nu\vee\delta u\rangle\, d\sigma + R(\theta,u) \tag{7.9}
$$

where $R(\theta,u)$ is another residual term which satisfies (7.2). Now (7.1) follows from an appropriate combination of the equalities (7.7), (7.8) and (7.9). ■

The usefulness of the identity contained in Proposition 7.1 is most apparent from the basic estimates we shall now deduce.

THEOREM 7.2. *Suppose that Ω is a bounded Lipschitz subdomain of $\mathcal{M}$ and $0\le l\le m$. Then for any l-form $u\in C^1(\overline{\Omega},\Lambda^l T\mathcal{M})$ such that $(\Delta - V)u = 0$ in Ω, we have*

$$
\begin{aligned}
&\max\left\{\begin{array}{l}\|\nu\wedge du\|_{L^2(\partial\Omega,\Lambda^{l+2}T\mathcal{M})}+\|\nu\wedge\delta u\|_{L^2(\partial\Omega,\Lambda^{l}T\mathcal{M})},\\[2ex] \|\nu\vee du\|_{L^2(\partial\Omega,\Lambda^{l}T\mathcal{M})}+\|\nu\vee\delta u\|_{L^2(\partial\Omega,\Lambda^{l-2}T\mathcal{M})}\end{array}\right\}\\
&\le C\min\left\{\begin{array}{l}\|\nu\wedge du\|_{L^2(\partial\Omega,\Lambda^{l+2}T\mathcal{M})}+\|\nu\wedge\delta u\|_{L^2(\partial\Omega,\Lambda^{l}T\mathcal{M})},\\[2ex] \|\nu\vee du\|_{L^2(\partial\Omega,\Lambda^{l}T\mathcal{M})}+\|\nu\vee\delta u\|_{L^2(\partial\Omega,\Lambda^{l-2}T\mathcal{M})}\end{array}\right\}\\
&\quad + C\left(\iint_{\Omega}(V^2|u|^2+|du|^2+|\delta u|^2)\, d\,Vol\right)^{1/2}
\end{aligned} \tag{7.10}
$$

uniformly in u and V.

PROOF. Let $\theta\in C^1(\mathcal{M},\Lambda^1 T\mathcal{M})$ have the property that $\langle\theta,\nu\rangle\ge\kappa>0$ a.e. on $\partial\Omega$. Proposition 7.1 gives

$$
\begin{aligned}
&\frac{\kappa}{2}\int_{\partial\Omega}|du|^2 d\sigma + \frac{\kappa}{2}\int_{\partial\Omega}|\delta u|^2 d\sigma\\
&\quad\le \int_{\partial\Omega}\tfrac{1}{2}|du|^2\langle\theta,\nu\rangle\, d\sigma + \int_{\partial\Omega}\tfrac{1}{2}|\delta u|^2\langle\theta,\nu\rangle\, d\sigma\\
&\quad\le \int_{\partial\Omega}\langle\theta\vee du,\nu\vee du\rangle\, d\sigma + \int_{\partial\Omega}\langle\theta\vee\delta u,\nu\vee\delta u\rangle\, d\sigma\\
&\qquad - \int_{\partial\Omega}\langle\nu\wedge\delta u,\theta\vee du\rangle\, d\sigma + R(u),
\end{aligned} \tag{7.11}
$$

where the error $R(u)$ can be controlled as follows

$$
|R(u)|\le C\left(\|Vu\|^2_{L^2(\Omega,\Lambda^l T\mathcal{M})}+\|du\|^2_{L^2(\Omega,\Lambda^{l+1}T\mathcal{M})}+\|\delta u\|^2_{L^2(\Omega,\Lambda^{l-1}T\mathcal{M})}\right). \tag{7.12}
$$

Further, we observe that $\langle \nu \wedge \delta u, \theta \vee du\rangle = -\langle \theta \wedge \delta u, \nu \vee du\rangle$. Then, for each fixed $\varepsilon > 0$, we have

$$\begin{aligned}
&\left| \int_{\partial\Omega} \langle \theta \vee du, \nu \vee du\rangle + \langle \theta \vee \delta u, \nu \vee \delta u\rangle + \langle \theta \wedge \delta u, \nu \vee du\rangle \, d\sigma \right| \\
&\leq \varepsilon \, \|du\|^2_{L^2(\partial\Omega, \Lambda^{l+1}T\mathcal{M})} + \frac{C}{\varepsilon}\|\nu \vee du\|^2_{L^2(\partial\Omega,\Lambda^l T\mathcal{M})} \\
&\quad + \varepsilon \, \|\delta u\|^2_{L^2(\partial\Omega,\Lambda^{l-1}T\mathcal{M})} + \frac{C}{\varepsilon}\|\nu \vee \delta u\|^2_{L^2(\partial\Omega,\Lambda^{l-2}T\mathcal{M})}
\end{aligned} \tag{7.13}$$

Choosing $\varepsilon \in (0, \frac{\kappa}{4})$ and absorbing the terms with small coefficients into the left side of (7.11) proves that the left-side of (7.10) is bounded by

$$\begin{aligned}
&C\|\nu \wedge du\|^2_{L^2(\partial\Omega,\Lambda^{l+2}T\mathcal{M})} + C\|\nu \wedge \delta u\|^2_{L^2(\partial\Omega,\Lambda^l T\mathcal{M})} \\
&+ C \iint_\Omega (V^2|u|^2 + |du|^2 + |\delta u|^2) \, d\mathrm{Vol}
\end{aligned}$$

If we now combine this with the similar estimate obtained by applying the same reasoning but to $*u$ in place of u, the conclusion follows. ■

Remark. In order to put matters in perspective, let us point out that the results above can be regarded as significant generalizations of the Rellich estimates for *scalar-valued* harmonic functions in Euclidean domains used by, e.g., D. Jerison, C. Kenig [JK1] and G. Verchota [Ve1]. Indeed, if $\mathcal{M}$ is the Euclidean space $\mathbb{R}^m$ and if u is a harmonic zero-form (i.e. a harmonic scalar-valued function) then $\delta u = 0$ and $du \equiv \nabla u$. Furthermore, $\|\nu \vee du\|_{L^2(\partial\Omega)} = \|\partial u/\partial\nu\|_{L^2(\partial\Omega)}$, and $\|\nu \wedge du\|_{L^2(\partial\Omega)} = \|\nabla_{\tan} u\|_{L^2(\partial\Omega)}$. Consequently, in this case the estimate in Theorem 7.2 takes the more familiar form

$$\|\nabla_{\tan} u\|_{L^2(\partial\Omega)} + \|u\|_{H^{1,2}(\Omega)} \approx \left\|\frac{\partial u}{\partial \nu}\right\|_{L^2(\partial\Omega)} + \|u\|_{H^{1,2}(\Omega)},$$

uniformly in u.

CHAPTER 8

Fredholm Properties of Boundary Integral Operators on Regular Spaces

With an eye toward proving the Fredholmness of the operators $\pm\frac{1}{2}I + M_l$ and $\pm\frac{1}{2}I + N_l$ on the spaces $L^p_{\tan}(\partial\Omega, \Lambda^l T\mathcal{M})$ and $L^p_{\rm nor}(\partial\Omega, \Lambda^l T\mathcal{M})$, respectively, in this chapter we will establish such results for the smaller spaces $L^{p,\delta}_{\tan}(\partial\Omega, \Lambda^l T\mathcal{M})$ and $L^{p,d}_{\rm nor}(\partial\Omega, \Lambda^l T\mathcal{M})$. Before embarking upon this, however, a few comments are in order. One basic ingredient in this analysis is deriving L^2-estimates to the effect that these operators are bounded from below modulo terms that we would like to treat as being residual. Specifically, let us pause to record the exact form of the estimate we have in mind.

PROPOSITION 8.1. *Assume that $\mathcal{M}$ is a smooth, oriented, compact manifold equipped with a metric tensor $g \in H^{2,r}(\mathcal{M}, Hom(T\mathcal{M} \otimes T\mathcal{M}, \mathbb{R}))$ for some $r > m = dim\,\mathcal{M}$. Also, let Ω be a bounded Lipschitz domain in $\mathcal{M}$ and assume that the scalar-valued, positive potential $V \in C^1(\mathcal{M})$, $V \neq 0$, satisfies $supp\,(dV) \cap \partial\Omega = \varnothing$.*

Then, for $\lambda \in \mathbb{R}$, $|\lambda| \geq \frac{1}{2}$, and $0 \leq l \leq m$, there exists $C = C(\partial\Omega, V, \lambda) > 0$ such that the following estimate holds for each $f \in L^2_{\tan}(\partial\Omega, \Lambda^l T\mathcal{M})$:

$$
\begin{aligned}
&\|f\|_{L^2(\partial\Omega,\Lambda^l T\mathcal{M})}\\
&\leq C\|(\lambda I + M_l)f\|_{L^2(\partial\Omega,\Lambda^l T\mathcal{M})} + C\|\nu \vee \delta \mathcal{S}_l f\|_{L^2(\partial\Omega,\Lambda^{l-2} T\mathcal{M})}\\
&\quad + C\|\mathcal{S}_l f\|_{H^{1,2}(\Omega_+,\Lambda^l T\mathcal{M})} + C\|\mathcal{S}_l f\|_{H^{1,2}(\Omega_-,\Lambda^l T\mathcal{M})}.
\end{aligned}
\tag{8.1}
$$

The proof is deferred to the second part of this chapter.

Before proceeding with the rigorous development we would like to emphasize one particular difficulty that arises in this context. In this vein, let us begin by pointing out that the term $\|\nu \vee \delta \mathcal{S}_l f\|_{L^2(\partial\Omega,\Lambda^{l-2} T\mathcal{M})}$ in the estimate (8.1) is zero if $l = 0$ or $l = 1$ (since $\nu \vee \delta \mathcal{S}_l f$ is a $(l-2)$-form if f is a l-form). This immediately leads to the conclusion that $\lambda I + M_l$ is bounded from below modulo compact operators on the "large" space $L^2_{\tan}(\partial\Omega, \Lambda^l T\mathcal{M})$, as desired, provided $l = 0$ or $l = 1$. However, for a general degree l, $\|\nu \vee \delta \mathcal{S}_l f\|_{L^2(\partial\Omega,\Lambda^{l-2} T\mathcal{M})}$ is a "big" term (the best upper bound in the general case is of the order of $\|f\|_{L^2(\partial\Omega,\Lambda^l T\mathcal{M})}$ which, of course, has no practical value in the present context), and this prevents us from inferring the same conclusion as directly as for small values of l. To circumvent this, we are thus forced to adopt the more circuitous route which has been outlined in some detail in the last part of § 6.

For the time being, in the next theorem we make the basic observation that in the case in which the operators are considered on spaces of *regular* forms then all residual terms appearing in the previous estimate are in fact *compact* operators.

THEOREM 8.2. *Assume the hypotheses of Proposition 8.1. Then, in the case when $f \in L^{2,\delta}_{\tan}(\partial\Omega, \Lambda^l T\mathcal{M})$, the following estimate holds:*

$$
\begin{aligned}
&\|f\|_{L^{2,\delta}_{\tan}(\partial\Omega,\Lambda^l T\mathcal{M})} \\
&\leq C\|(\lambda I + M_l)f\|_{L^{2,\delta}_{\tan}(\partial\Omega,\Lambda^l T\mathcal{M})} + C\|S_{l-1}(\delta_\partial f)\|_{L^2(\partial\Omega,\Lambda^{l-1}T\mathcal{M})} \\
&\quad + C\|\mathcal{S}_l f\|_{H^{1,2}(\Omega_+,\Lambda^l T\mathcal{M})} + C\|\mathcal{S}_{l-1}(\delta_\partial f)\|_{H^{1,2}(\Omega_+,\Lambda^{l-1}T\mathcal{M})} \\
&\quad + C\|\mathcal{S}_l f\|_{H^{1,2}(\Omega_-,\Lambda^l T\mathcal{M})} + C\|\mathcal{S}_{l-1}(\delta_\partial f)\|_{H^{1,2}(\Omega_-,\Lambda^{l-1}T\mathcal{M})} \\
&\quad + C\|\nu \vee R_{l-1}f\|_{L^2(\partial\Omega,\Lambda^l T\mathcal{M})} + C\|\nu \vee R_{l-2}(\delta_\partial f)\|_{L^2(\partial\Omega,\Lambda^{l-1}T\mathcal{M})}.
\end{aligned}
\tag{8.2}
$$

In particular, for any $\lambda \in \mathbb{R}$ with $|\lambda| \geq \frac{1}{2}$ the operator

$$\lambda I + M_l : L^{2,\delta}_{\tan}(\partial\Omega, \Lambda^l T\mathcal{M}) \to L^{2,\delta}_{\tan}(\partial\Omega, \Lambda^l T\mathcal{M}) \tag{8.3}$$

is Fredholm with index zero.

Similar conclusions are valid for $\lambda I + N_l$ acting on $L^{2,d}_{\mathrm{nor}}(\partial\Omega, \Lambda^l T\mathcal{M})$.

PROOF. Combining (the boundary trace version of) (6.17) with (8.1) yields

$$
\begin{aligned}
&\|f\|_{L^2(\partial\Omega,\Lambda^l T\mathcal{M})} \\
&\leq C\|(\lambda I + M_l)f\|_{L^2(\partial\Omega,\Lambda^l T\mathcal{M})} + C\|\nu \vee S_{l-1}(\delta_\partial f)\|_{L^2(\partial\Omega,\Lambda^{l-2}T\mathcal{M})} \\
&\quad + C\|\mathcal{S}_l f\|_{H^{1,2}(\Omega_+,\Lambda^l T\mathcal{M})} + C\|\mathcal{S}_l f\|_{H^{1,2}(\Omega_-,\Lambda^l T\mathcal{M})} \\
&\quad + C\|\nu \vee R_{l-1}f\|_{L^2(\partial\Omega,\Lambda^l T\mathcal{M})},
\end{aligned}
\tag{8.4}
$$

uniformly for $f \in L^{2,\delta}_{\tan}(\partial\Omega, \Lambda^l T\mathcal{M})$. The key observation now is that $\delta_\partial f$ belongs to $L^{2,0}_{\tan}(\partial\Omega, \Lambda^{l-1}T\mathcal{M})$ and, therefore, we can re-apply this inequality (with l replaced by $l-1$) to $\delta_\partial f$ in place of f. On account of the "almost" commutativity identity (6.19) this reads

$$
\begin{aligned}
\|\delta_\partial f\|_{L^2(\partial\Omega,\Lambda^{l-1}T\mathcal{M})} &\leq C\|\delta_\partial\left((\lambda I + M_l)f\right)\|_{L^2(\partial\Omega,\Lambda^{l-1}T\mathcal{M})} \\
&\quad + C\|\mathcal{S}_{l-1}(\delta_\partial f)\|_{H^{1,2}(\Omega_+,\Lambda^{l-1}T\mathcal{M})} \\
&\quad + C\|\mathcal{S}_{l-1}(\delta_\partial f)\|_{H^{1,2}(\Omega_-,\Lambda^{l-1}T\mathcal{M})} \\
&\quad + C\|\nu \vee R_{l-2}(\delta_\partial f)\|_{L^2(\partial\Omega,\Lambda^{l-1}T\mathcal{M})}.
\end{aligned}
\tag{8.5}
$$

The estimate (8.2) then follows by adding up (8.4) and (8.5).

What this estimate amounts to is the boundedness from below of the operator (8.3) modulo compact operators. Thus, the operator (8.3) is semi-Fredholm and has a well-defined index. Now, the index is an integer-valued function which depends continuously on the parameter $\lambda \in \mathbb{R} \setminus (-\frac{1}{2}, \frac{1}{2})$. Since it clearly vanishes for large $|\lambda|$, we infer that it vanishes identically for $\lambda \in \mathbb{R} \setminus (-\frac{1}{2}, \frac{1}{2})$. Consequently, the operator (8.3) is Fredholm with index zero. ∎

COROLLARY 8.3. *Assume the same hypotheses as in Proposition 8.1 and, in addition, suppose that the potential V is a strictly positive constant. Then the eigenvalues of the operators M_l on $L^{2,\delta}_{\tan}(\partial\Omega, \Lambda^l T\mathcal{M})$ and N_l on $L^{2,d}_{\mathrm{nor}}(\partial\Omega, \Lambda^l T\mathcal{M})$ are contained in the interval $(-\frac{1}{2}, \frac{1}{2})$.*

In particular, the operators

$$\lambda I + M_l : L^{2,\delta}_{\tan}(\partial\Omega, \Lambda^l T\mathcal{M}) \to L^{2,\delta}_{\tan}(\partial\Omega, \Lambda^l T\mathcal{M}) \tag{8.6}$$

and

$$\lambda I + N_l : L^{2,d}_{\mathrm{nor}}(\partial\Omega, \Lambda^l T\mathcal{M}) \to L^{2,d}_{\mathrm{nor}}(\partial\Omega, \Lambda^l T\mathcal{M}) \tag{8.7}$$

are invertible for any $\lambda \in \mathbb{R}$ *with* $|\lambda| \geq \frac{1}{2}$.

PROOF. Let $z \in \mathbb{C}$ be such that there exists $f \in L^{2,\delta}_{\mathrm{tan}}(\partial\Omega, \Lambda^l T\mathcal{M})$, $f \neq 0$, with $M_l f = zf$ and set $u := d\mathcal{S}_l f$ in $\mathcal{M} \setminus \partial\Omega$. Then, since $[d, V] = 0$ in this case, we have $(\Delta - V)u = 0$ in $\mathcal{M} \setminus \partial\Omega$. This and the fact that u has appropriate control at the boundary (i.e., $\mathcal{N}u$, $\mathcal{N}(du)$, $\mathcal{N}(\delta u) \in L^2(\partial\Omega_\pm)$) can be used to prove, via a limiting argument and repeated integrations by parts, the following energy identities

$$\begin{aligned}\iint_{\Omega_\pm} \left(|du|^2 + |\delta u|^2 + V|u|^2\right) d\mathrm{Vol} &= \mp \int_{\partial\Omega_\pm} \langle \nu \vee u,\, \nu \vee (\nu \wedge \delta\bar{u})\rangle \, d\sigma \\ &\quad \pm \int_{\partial\Omega_\pm} \langle \nu \wedge u,\, \nu \wedge (\nu \vee d\bar{u})\rangle \, d\sigma.\end{aligned} \tag{8.8}$$

Denote by $\mu_\pm \in \mathbb{R}$, respectively, the solid integrals in the left side of (8.8). In our case, $du = 0$ in $\mathcal{M} \setminus \partial\Omega$, $\nu \vee u|_{\partial\Omega_\pm} = (\mp\frac{1}{2} + z)f$ and $\nu \wedge (\delta u)|_{\partial\Omega_+} = \nu \wedge (\delta u)|_{\partial\Omega_-}$. Using these, we see that

$$\mu_\pm = \left(\tfrac{1}{2} \mp z\right) \int_{\partial\Omega} \langle f, \nu \vee (\nu \wedge \delta\bar{u})\rangle \, d\sigma.$$

Unless either μ_+ or μ_- vanishes, in which case it is easy to see (from (8.8) and jump-relations) that f must be zero, it follows that $z = -(\mu_+ - \mu_-)/2(\mu_+ + \mu_-) \in \mathbb{R}$ and

$$|z| = \frac{1}{2}\left|\frac{\mu_+ - \mu_-}{\mu_+ + \mu_-}\right| < \frac{1}{2},$$

as desired. A similar reasoning applies to the operator N_l (alternatively, one may use the identity (6.12)) and this proves the first part of the corollary.

The second part of the corollary follows from what we have proved so far and Theorem 8.2. ■

The second part of this chapter is devoted to presenting the

PROOF OF PROPOSITION 8.1. First we consider the cases when $\lambda = \pm\frac{1}{2}$. For a fixed, arbitrary $f \in L^2_{\mathrm{tan}}(\partial\Omega, \Lambda^l T\mathcal{M})$ using (6.8) and the tangentiality of f it follows that $(\delta\mathcal{S}_l f)|_{\partial\Omega_\pm} = \delta\mathcal{S}_l f$ and, consequently, $\nu \wedge (\delta\mathcal{S}_l f)|_{\partial\Omega_\pm} = \nu \wedge \delta\mathcal{S}_l f$, and $\nu \vee (\delta\mathcal{S}_l f)|_{\partial\Omega_\pm} = \nu \vee \delta\mathcal{S}_l f$. Similarly, $\nu \wedge (d\mathcal{S}_l f)|_{\partial\Omega_\pm} = \nu \wedge d\mathcal{S}_l f$ and recall from the previous chapter that $\nu \vee (d\mathcal{S}_l f)|_{\partial\Omega_\pm} = \left(\mp\frac{1}{2}I + M_l\right) f$.

Next, we utilize the estimate in Theorem 7.2 for Ω_- and $u := \mathcal{S}_l f$, and replace the boundary traces according to the trace formulas presented above. This and the triangle inequality then give

$$\begin{aligned}&\|f\|_{L^2(\partial\Omega,\Lambda^l T\mathcal{M})} \\ &\leq \left\|(-\tfrac{1}{2}I + M_l)f\right\|_{L^2(\partial\Omega,\Lambda^l T\mathcal{M})} + \left\|(\tfrac{1}{2}I + M_l)f\right\|_{L^2(\partial\Omega,\Lambda^l T\mathcal{M})} \\ &= \left\|(-\tfrac{1}{2}I + M_l)f\right\|_{L^2(\partial\Omega,\Lambda^l T\mathcal{M})} + \left\|\nu \vee (d\mathcal{S}_l f)|_{\partial\Omega_-}\right\|_{L^2(\partial\Omega,\Lambda^l T\mathcal{M})} \\ &\leq \left\|(-\tfrac{1}{2}I + M_l)f\right\|_{L^2(\partial\Omega,\Lambda^l T\mathcal{M})} + C\left\|\nu \wedge (d\mathcal{S}_l f)|_{\partial\Omega_-}\right\|_{L^2(\partial\Omega,\Lambda^{l+2} T\mathcal{M})} \\ &\quad + C\left\|\nu \wedge (\delta\mathcal{S}_l f)|_{\partial\Omega_-}\right\|_{L^2(\partial\Omega,\Lambda^l T\mathcal{M})} + C\left\|\mathcal{S}_l f\right\|_{H^{1,2}(\Omega_-,\Lambda^l T\mathcal{M})}.\end{aligned}$$

We see from the trace formulas recorded above that $\nu\wedge(d\mathcal{S}_l f)$ and $\nu\wedge(\delta\mathcal{S}_l f)$ have no jump across $\partial\Omega$, hence, actually

$$\begin{aligned}
&\|f\|_{L^2(\partial\Omega,\Lambda^l T\mathcal{M})}\\
&\le \big\|(-\tfrac{1}{2}I+M_l)f\big\|_{L^2(\partial\Omega,\Lambda^l T\mathcal{M})}+C\big\|\nu\wedge(d\mathcal{S}_l f)|_{\partial\Omega_+}\big\|_{L^2(\partial\Omega,\Lambda^{l+2}T\mathcal{M})}\\
&\quad +C\big\|\nu\wedge(\delta\mathcal{S}_l f)|_{\partial\Omega_+}\big\|_{L^2(\partial\Omega,\Lambda^l T\mathcal{M})}+C\big\|\mathcal{S}_l f\big\|_{H^{1,2}(\Omega_-,\Lambda^l T\mathcal{M})}.
\end{aligned}$$

Applying once again (7.10), written for the same u as before although, this time, considered in Ω_+, allows us to control the traces on $\partial\Omega_+$ in the previous estimate so we finally arrive at

$$\begin{aligned}
&\|f\|_{L^2(\partial\Omega,\Lambda^l T\mathcal{M})}\\
&\le \big\|(-\tfrac{1}{2}I+M_l)f\big\|_{L^2(\partial\Omega,\Lambda^l T\mathcal{M})}+C\big\|\nu\vee(d\mathcal{S}_l f)|_{\partial\Omega_+}\big\|_{L^2(\partial\Omega,\Lambda^l T\mathcal{M})}\\
&\quad +C\big\|\nu\vee(\delta\mathcal{S}_l f)|_{\partial\Omega_+}\big\|_{L^2(\partial\Omega,\Lambda^{l-2}T\mathcal{M})}+C\big\|\mathcal{S}_l f\big\|_{H^{1,2}(\Omega_+,\Lambda^l T\mathcal{M})}\\
&\quad +C\big\|\mathcal{S}_l f\big\|_{H^{1,2}(\Omega_-,\Lambda^l T\mathcal{M})}\\
&= C\big\|(-\tfrac{1}{2}I+M_l)f\big\|_{L^2(\partial\Omega,\Lambda^l T\mathcal{M})}+C\big\|\nu\vee(\delta\mathcal{S}_l f)|_{\partial\Omega_+}\big\|_{L^2(\partial\Omega,\Lambda^{l-2}T\mathcal{M})}\\
&\quad +C\big\|\mathcal{S}_l f\big\|_{H^{1,2}(\Omega_+,\Lambda^l T\mathcal{M})}+C\big\|\mathcal{S}_l f\big\|_{H^{1,2}(\Omega_-,\Lambda^l T\mathcal{M})}.
\end{aligned}$$

This proves the desired estimate for $\lambda=-\frac{1}{2}$. The arguments for $\lambda=+\frac{1}{2}$ are essentially a repetition of the above ones and, hence, are omitted.

Handling the case $|\lambda|>\frac{1}{2}$ is somewhat more involved. To begin with, we take the identity (7.1) in Proposition 7.1 written for $\Omega=\Omega_+$, $u=\mathcal{S}_l f$ and multiply it by $\lambda-\frac{1}{2}$. Keeping the same u, we then take the identity (7.1), written this time for Ω_- in place of Ω, and multiply it by $\lambda+\frac{1}{2}$. Summing the two gives

$$\begin{aligned}
&\big(\lambda-\tfrac{1}{2}\big)\int_{\partial\Omega_+}\langle\theta\vee du,\nu\vee du\rangle\,d\sigma-\big(\lambda+\tfrac{1}{2}\big)\int_{\partial\Omega_-}\langle\theta\vee du,\nu\vee du\rangle\,d\sigma\\
&\big(\lambda-\tfrac{1}{2}\big)\int_{\partial\Omega_+}\langle\theta\vee \delta u,\nu\vee \delta u\rangle\,d\sigma-\big(\lambda+\tfrac{1}{2}\big)\int_{\partial\Omega_-}\langle\theta\vee \delta u,\nu\vee \delta u\rangle\,d\sigma\\
&-\big(\lambda-\tfrac{1}{2}\big)\int_{\partial\Omega_+}\langle\nu\wedge\delta u,\theta\vee du\rangle\,d\sigma+\big(\lambda+\tfrac{1}{2}\big)\int_{\partial\Omega_-}\langle\nu\wedge\delta u,\theta\vee du\rangle\,d\sigma\\
&=\big(\lambda-\tfrac{1}{2}\big)\int_{\partial\Omega_+}\tfrac{1}{2}|du|^2\langle\theta,\nu\rangle\,d\sigma-\big(\lambda+\tfrac{1}{2}\big)\int_{\partial\Omega_-}\tfrac{1}{2}|du|^2\langle\theta,\nu\rangle\,d\sigma\\
&\quad+\big(\lambda-\tfrac{1}{2}\big)\int_{\partial\Omega_+}\tfrac{1}{2}|\delta u|^2\langle\theta,\nu\rangle\,d\sigma-\big(\lambda+\tfrac{1}{2}\big)\int_{\partial\Omega_-}\tfrac{1}{2}|\delta u|^2\langle\theta,\nu\rangle\,d\sigma\\
&\quad+\mathcal{O}\Big(\|u\|^2_{H^{1,2}(\Omega_+,\Lambda^l T\mathcal{M})}+\|u\|^2_{H^{1,2}(\Omega_-,\Lambda^l T\mathcal{M})}\Big).
\end{aligned}\tag{8.9}$$

Next, we shall analyze the boundary integrals in the above identity. For the first line of (8.9) we have

$$
\begin{aligned}
1^{\text{st}} \text{ line} &= \left(\lambda - \tfrac{1}{2}\right) \int_{\partial\Omega} \langle \theta \vee (du)|_{\partial\Omega_+}, -\left(\tfrac{1}{2} + \lambda\right) f + (\lambda I + M_l) f \rangle \, d\sigma \\
&\quad - \left(\lambda + \tfrac{1}{2}\right) \int_{\partial\Omega} \langle \theta \vee (du)|_{\partial\Omega_-}, -\left(\lambda - \tfrac{1}{2}\right) f + (\lambda I + M_l) f \rangle \, d\sigma \\
&= \left(\lambda - \tfrac{1}{2}\right) \int_{\partial\Omega} \langle \theta \vee (du)|_{\partial\Omega_+}, (\lambda I + M_l) f \rangle \, d\sigma \\
&\quad - \left(\lambda + \tfrac{1}{2}\right) \int_{\partial\Omega} \langle \theta \vee (du)|_{\partial\Omega_-}, (\lambda I + M_l) f \rangle \, d\sigma \\
&\quad + \left(\lambda^2 - \tfrac{1}{4}\right) \int_{\partial\Omega} \langle \theta \vee ((du)|_{\partial\Omega_-} - (du)|_{\partial\Omega_+}), f \rangle \, d\sigma.
\end{aligned} \tag{8.10}
$$

In the last integrand above we use the tangentiality of f and Lemma 4.1 to write

$$
\begin{aligned}
\langle \theta \vee ((du)|_{\partial\Omega_-} - (du)|_{\partial\Omega_+}), f \rangle &= \langle \theta \vee (\nu \wedge f), f \rangle = \langle f, \nu \vee (\theta \wedge f) \rangle \\
&= \langle f, \langle \theta, \nu \rangle f \rangle - \langle f, \theta \wedge (\nu \vee f) \rangle \\
&= |f|^2 \langle \theta, \nu \rangle.
\end{aligned} \tag{8.11}
$$

For the second and the fifth line in the identity (8.9) the terms under the integral signs do not jump across $\partial\Omega$. Their difference can be seen immediately to be

$$
2^{\text{nd}} \text{ line} - 5^{\text{th}} \text{ line} = \tfrac{1}{2} \int_{\partial\Omega} |\delta S_l f|^2 \langle \theta, \nu \rangle \, d\sigma - \int_{\partial\Omega} \langle \theta \vee \delta S_l f \, , \, \nu \vee \delta S_l f \rangle \, d\sigma. \tag{8.12}
$$

Going further, the third line of the identity (8.9) becomes

$$
\begin{aligned}
3^{\text{rd}} \text{ line} &= -\left(\lambda - \tfrac{1}{2}\right) \int_{\partial\Omega} \langle \nu \wedge \delta S_l f \, , \theta \vee \left(-\tfrac{1}{2} \nu \wedge f + d S_l f\right) \rangle \, d\sigma \\
&\quad + \left(\lambda + \tfrac{1}{2}\right) \int_{\partial\Omega} \langle \nu \wedge \delta S_l f \, , \theta \vee \left(\tfrac{1}{2} \nu \wedge f + d S_l f\right) \rangle \, d\sigma \\
&= \left(\lambda - \tfrac{1}{2}\right) \int_{\partial\Omega} \langle \theta \wedge \delta S_l f \, , \, \nu \vee \left(-\tfrac{1}{2} \nu \wedge f + d S_l f\right) \rangle \, d\sigma \\
&\quad - \left(\lambda + \tfrac{1}{2}\right) \int_{\partial\Omega} \langle \theta \wedge \delta S_l f \, , \, \nu \vee \left(\tfrac{1}{2} \nu \wedge f + d S_l f\right) \rangle \, d\sigma \\
&= \left(\lambda - \tfrac{1}{2}\right) \int_{\partial\Omega} \langle \theta \wedge \delta S_l f \, , \, \left(-\tfrac{1}{2} + \lambda\right) f + (\lambda I + M_l) f \rangle \, d\sigma \\
&\quad - \left(\lambda + \tfrac{1}{2}\right) \int_{\partial\Omega} \langle \theta \wedge \delta S_l f \, , \, \left(\tfrac{1}{2} - \lambda\right) f + (\lambda I + M_l) f \rangle \, d\sigma \\
&= - \int_{\partial\Omega} \langle \theta \wedge \delta S_l f \, , \, (\lambda I + M_l) f \rangle \, d\sigma.
\end{aligned} \tag{8.13}
$$

Next, since $|du|^2 = |\nu \wedge du|^2 + |\nu \vee du|^2$, after replacing du on $\partial\Omega_\pm$ with the corresponding non-tangential boundary limits, we see that the fourth line in (8.9) can be rewritten in the form

$$
\begin{aligned}
4^{\text{th}}\text{ line} &= \left(\lambda - \tfrac{1}{2}\right)\int_{\partial\Omega_+} \tfrac{1}{2}|du|^2\langle\theta,\nu\rangle\,d\sigma - \left(\lambda + \tfrac{1}{2}\right)\int_{\partial\Omega_-} \tfrac{1}{2}|du|^2\langle\theta,\nu\rangle\,d\sigma \\
&= -\tfrac{1}{2}\int_{\partial\Omega} |\nu\wedge dS_l f|^2\langle\theta,\nu\rangle\,d\sigma + \tfrac{1}{2}\left(\lambda - \tfrac{1}{2}\right)\int_{\partial\Omega} \left|\left(-\tfrac{1}{2}I + M_l\right)f\right|^2\langle\theta,\nu\rangle\,d\sigma \\
&\quad - \tfrac{1}{2}\left(\lambda + \tfrac{1}{2}\right)\int_{\partial\Omega} \left|\left(\tfrac{1}{2}I + M_l\right)f\right|^2\langle\theta,\nu\rangle\,d\sigma.
\end{aligned} \tag{8.14}
$$

In the last two integrals above we express $\pm\frac{1}{2}I + M_l$ as $(\pm\frac{1}{2} - \lambda)I + (\lambda I + M_l)$. Consequently, a straightforward calculation shows that

$$
\begin{aligned}
4^{\text{th}}\text{ line} &= -\tfrac{1}{2}\int_{\partial\Omega} |\nu\wedge dS_l f|^2\langle\theta,\nu\rangle\,d\sigma \\
&\quad + \tfrac{1}{2}\left(\lambda - \tfrac{1}{2}\right)\int_{\partial\Omega} \left|\left(-\tfrac{1}{2}I + M_l\right)f\right|^2\langle\theta,\nu\rangle\,d\sigma \\
&\quad - \tfrac{1}{2}\left(\lambda + \tfrac{1}{2}\right)\int_{\partial\Omega} \left|\left(\tfrac{1}{2}I + M_l\right)f\right|^2\langle\theta,\nu\rangle\,d\sigma \\
&= -\tfrac{1}{2}\int_{\partial\Omega} |\nu\wedge dS_l f|^2\langle\theta,\nu\rangle\,d\sigma + \tfrac{1}{2}\left(\lambda^2 - \tfrac{1}{4}\right)\int_{\partial\Omega} |f|^2\langle\theta,\nu\rangle\,d\sigma \\
&\quad - \tfrac{1}{2}\int_{\partial\Omega} |(\lambda I + M_l)f|^2\langle\theta,\nu\rangle\,d\sigma.
\end{aligned} \tag{8.15}
$$

At this point, combining the identities (8.9)–(8.15) we see that

$$
\begin{aligned}
&\tfrac{1}{2}\left(\lambda^2 - \tfrac{1}{4}\right)\int_{\partial\Omega} |f|^2\langle\theta,\nu\rangle\,d\sigma + \tfrac{1}{2}\int_{\partial\Omega} |(\lambda I + M_l)f|^2\langle\theta,\nu\rangle\,d\sigma \\
&\quad + \tfrac{1}{2}\int_{\partial\Omega} |\nu\wedge dS_l f|^2\langle\theta,\nu\rangle\,d\sigma + \tfrac{1}{2}\int_{\partial\Omega} |\delta u|^2\langle\theta,\nu\rangle\,d\sigma \\
&= -\left(\lambda - \tfrac{1}{2}\right)\int_{\partial\Omega} \langle\theta\vee(dS_l f)|_{\partial\Omega_+}, (\lambda I + M_l)f\rangle\,d\sigma \\
&\quad + \left(\lambda + \tfrac{1}{2}\right)\int_{\partial\Omega} \langle\theta\vee(dS_l f)|_{\partial\Omega_-}, (\lambda I + M_l)f\rangle\,d\sigma \\
&\quad + \int_{\partial\Omega} \langle\theta\wedge\delta S_l f\,, (\lambda I + M_l)f\rangle\,d\sigma + \int_{\partial\Omega} \langle\theta\vee\delta S_l f\,, \nu\vee\delta S_l f\rangle\,d\sigma \\
&\quad + \mathcal{O}\left(\|S_l f\|^2_{H^{1,2}(\Omega_+,\Lambda^l T\mathcal{M})} + \|S_l f\|^2_{H^{1,2}(\Omega_-,\Lambda^l T\mathcal{M})}\right).
\end{aligned} \tag{8.16}
$$

Note that, since matters have been arranged so that $\langle\theta,\nu\rangle \geq \kappa > 0$, the left side of (8.16) dominates $\frac{\kappa}{2}(\lambda^2 - \frac{1}{4})\|f\|^2_{L^2(\partial\Omega,\Lambda^l T\mathcal{M})}$ simply because of sign considerations. Turning attention to the right side, first, by repeated applications of Schwarz's inequality and the boundedness of the boundary integral operators dS_l, δS_l, we obtain the estimate

$$\begin{aligned}\Big|\int_{\partial\Omega}\Big[\langle\theta\vee(d\mathcal{S}_l f)|_{\partial\Omega_\pm},(\lambda I+M_l)f\rangle+\langle\theta\wedge\delta\mathcal{S}_l f,(\lambda I+M_l)f\rangle\\ +\langle\theta\vee\delta\mathcal{S}_l f,\nu\vee\delta\mathcal{S}_l f\rangle\Big]\,d\sigma\Big|\\ \leq\varepsilon\,\|f\|^2_{L^2(\partial\Omega,\Lambda^l T\mathcal{M})}+\frac{C}{\varepsilon}\|(\lambda I+M_l)f\|^2_{L^2(\partial\Omega,\Lambda^l T\mathcal{M})}\\ +\frac{C}{\varepsilon}\|\nu\vee\delta\mathcal{S}_l f\|^2_{L^2(\partial\Omega,\Lambda^{l-2}T\mathcal{M})},\end{aligned}$$

valid for any $\varepsilon > 0$.

In sum, we deduce from (8.16) that

$$\begin{aligned}&\tfrac{1}{2}\big(\lambda^2-\tfrac{1}{4}\big)\int_{\partial\Omega}|f|^2\langle\theta,\nu\rangle\,d\sigma\\ &\leq\varepsilon\,\|f\|^2_{L^2(\partial\Omega,\Lambda^l T\mathcal{M})}+\frac{C}{\varepsilon}\|(\lambda I+M_l)f\|^2_{L^2(\partial\Omega,\Lambda^l T\mathcal{M})}\\ &\quad+C\|\mathcal{S}_l f\|^2_{H^{1,2}(\Omega_+,\Lambda^l T\mathcal{M})}+C\|\mathcal{S}_l f\|^2_{H^{1,2}(\Omega,\Lambda^l T\mathcal{M})}\\ &\quad+C\|\mathcal{S}_l f\|^2_{H^{1,2}(\Omega_-,\Lambda^l T\mathcal{M})}+\frac{C}{\varepsilon}\|\nu\vee\delta\mathcal{S}_l f\|^2_{L^2(\partial\Omega,\Lambda^l T\mathcal{M})}.\end{aligned}\tag{8.17}$$

Finally, since $|\lambda| > \frac{1}{2}$, we can choose ε small enough such that $\lambda^2 - \frac{1}{4} - \frac{\varepsilon}{2} > 0$. Then, the norm estimate stated in Proposition 8.1 follows and the proof of that result is complete. ∎

CHAPTER 9

Weak Extensions of Boundary Derivative Operators

We retain the hypotheses on $\mathcal{M}$ and Ω from the previous chapter (cf. the statement of Proposition 8.1). In order to set the stage for the main result of this chapter, we debut with presenting an integral identity, which will play an important role in the sequel.

PROPOSITION 9.1. *For each $l = 0, ..., m$ and any forms $f \in L^{p,\delta}_{\tan}(\partial\Omega, \Lambda^l T\mathcal{M})$ and $g \in L^{q,d}_{\rm nor}(\partial\Omega, \Lambda^l T\mathcal{M})$ with $1/p + 1/q = 1$, the following formula holds:*

$$\int_{\partial\Omega} \langle \delta_\partial f \,,\, \nu \vee g \rangle \, d\sigma = - \int_{\partial\Omega} \langle \nu \wedge f \,,\, d_\partial g \rangle \, d\sigma. \tag{9.1}$$

PROOF. It suffices to treat the case $p = q = 2$ since, as it follows from, e.g., the results proved in Appendix B, $L^{p,\delta}_{\tan}(\partial\Omega, \Lambda^l T\mathcal{M}) \cap L^{2,\delta}_{\tan}(\partial\Omega, \Lambda^l T\mathcal{M})$ is dense in $L^{p,\delta}_{\tan}(\partial\Omega, \Lambda^l T\mathcal{M})$ for any $1 < p < \infty$, plus a similar property for normal forms.

To this end, let now V be a constant, positive potential so that, by invoking Corollary 8.3, we may take

$$f' := (-\tfrac{1}{2}I + M_l)^{-1} f \in L^{2,\delta}_{\tan}(\partial\Omega, \Lambda^l T\mathcal{M}), \quad g' := (\tfrac{1}{2}I + N_l)^{-1} g \in L^{2,d}_{\rm nor}(\partial\Omega, \Lambda^l T\mathcal{M}).$$

If we now set $u := d\mathcal{S}_l f'$ and $\omega := \delta\mathcal{S}_l g'$ in Ω, then $f = \nu \vee u|_{\partial\Omega}$ and $g = \nu \wedge \omega|_{\partial\Omega}$. Moreover, since u, ω also have appropriate control at the boundary, by (6.15)–(6.16), we have $\delta_\partial f = -\nu \vee (\delta u)|_{\partial\Omega}$ and $d_\partial g = -\nu \wedge (d\omega)|_{\partial\Omega}$. With these at hand and using integration by parts we may thus write

$$\begin{aligned}
&\int_{\partial\Omega} \langle \delta_\partial f \,,\, \nu \vee g \rangle \, d\sigma \\
&\quad = \int_{\partial\Omega} \langle -\nu \vee \delta u, \, \nu \vee g \rangle \, d\sigma = - \int_{\partial\Omega} \langle \delta u, \, \nu \wedge (\nu \vee (\nu \wedge \omega)) \rangle \, d\sigma \\
&\quad = - \int_{\partial\Omega} \langle \delta u, \nu \wedge \omega \rangle \, d\sigma = - \iint_{\Omega} \langle \delta u \,,\, d\omega \rangle \, d\mathrm{Vol} = \int_{\partial\Omega} \langle \nu \vee u \,,\, d\omega \rangle \, d\sigma \\
&\quad = - \int_{\partial\Omega} \langle \nu \wedge (\nu \vee u) \,, -\nu \wedge d\omega \rangle \, d\sigma = - \int_{\partial\Omega} \langle \nu \wedge f \,,\, d_\partial g \rangle \, d\sigma
\end{aligned}$$

and the conclusion follows. ∎

The identity (9.1) suggests the possibility of extending the action of the boundary derivative operator δ_∂, originally mapping its "natural" domain $L^{p,\delta}_{\tan}(\partial\Omega, \Lambda^l T\mathcal{M})$

into $\delta_\partial L^{p,\delta}_{\rm tan}(\partial\Omega,\Lambda^l T\mathcal{M}) \subseteq L^{p,0}_{\rm tan}(\partial\Omega,\Lambda^{l-1}T\mathcal{M})$, to a larger space. Specifically, we shall consider the *weak* extension

$$\delta_\partial : L^p_{\rm tan}(\partial\Omega,\Lambda^l T\mathcal{M}) \to \left(\frac{\nu\vee L^{q,d}_{\rm nor}(\partial\Omega,\Lambda^l T\mathcal{M})}{\nu\vee L^{q,0}_{\rm nor}(\partial\Omega,\Lambda^l T\mathcal{M})}\right)^*, \quad \frac{1}{p}+\frac{1}{q}=1, \tag{9.2}$$

defined by

$$\langle\!\langle \delta_\partial f, [\nu\vee g]\rangle\!\rangle := -\int_{\partial\Omega}\langle \nu\wedge f, d_\partial g\rangle\, d\sigma \tag{9.3}$$

for any $f\in L^p_{\rm tan}(\partial\Omega,\Lambda^l T\mathcal{M})$ and $g\in L^{q,d}_{\rm nor}(\partial\Omega,\Lambda^l T\mathcal{M})$. Here $\langle\!\langle\cdot,\cdot\rangle\!\rangle$ stands for the natural duality pairing, whereas $[\cdot]$ denotes the projection operator onto the quotient space under discussion. Also, the star superscript indicates the dual space. It is not too difficult to check that, under an appropriate embedding

$$\delta_\partial L^{p,\delta}_{\rm tan}(\partial\Omega,\Lambda^l T\mathcal{M}) \hookrightarrow \left(\frac{\nu\vee L^{q,d}_{\rm nor}(\partial\Omega,\Lambda^l T\mathcal{M})}{\nu\vee L^{q,0}_{\rm nor}(\partial\Omega,\Lambda^l T\mathcal{M})}\right)^*,$$

and, as such, the operator in (9.3) is a *genuine* extension of the "old" δ_∂ (introduced in (5.1)). In the sequel, we shall frequently use this fact without any special mention. Note that this also justifies our keeping the same notation for both operators.

We are now in a position to state the main result of this chapter.

THEOREM 9.2. *For any $1<p,q<\infty$, $1/p+1/q=1$ and each $l\in\{0,1,\dots,m\}$ the operator*

$$\delta_\partial : \frac{L^p_{\rm tan}(\partial\Omega,\Lambda^l T\mathcal{M})}{L^{p,0}_{\rm tan}(\partial\Omega,\Lambda^l T\mathcal{M})} \to \left(\frac{\nu\vee L^{q,d}_{\rm nor}(\partial\Omega,\Lambda^l T\mathcal{M})}{\nu\vee L^{q,0}_{\rm nor}(\partial\Omega,\Lambda^l T\mathcal{M})}\right)^* \tag{9.4}$$

is an isomorphism.

Of course, a similar result is valid for d_∂ also. The proof of Theorem 9.2 makes use of the following calculation of the index of d_∂.

PROPOSITION 9.3. *For any $1<p<\infty$, the operator*

$$d_\partial : L^{p,d}_{\rm nor}(\partial\Omega,\Lambda^l T\mathcal{M}) \to L^{p,0}_{\rm nor}(\partial\Omega,\Lambda^{l+1} T\mathcal{M})$$

has closed range and its cokernel is isomorphic to $H^l_{\rm sing}(\partial\Omega;\mathbb{R})$, the l-th singular homology group of $\partial\Omega$ over the reals. In particular,

$$d_\partial : \frac{L^{p,d}_{\rm nor}(\partial\Omega,\Lambda^l T\mathcal{M})}{L^{p,0}_{\rm nor}(\partial\Omega,\Lambda^l T\mathcal{M})} \to L^{p,0}_{\rm nor}(\partial\Omega,\Lambda^{l+1} T\mathcal{M}) \tag{9.5}$$

is a Fredholm operator with index $b_l(\partial\Omega)$, the l-th Betti number of $\partial\Omega$.

We defer the proof of this result to the second part of this chapter and at this point present the

PROOF OF THEOREM 9.2. By the definition of δ_∂ and the discussion above, the operator in (9.4) is one-to-one. Next we check that it is also onto. To this end, we first note that its range does not change if the original domain is replaced by

$$\frac{L^p_{\rm tan}(\partial\Omega,\Lambda^l T\mathcal{M})}{\delta_\partial(L^{p,\delta}_{\rm tan}(\partial\Omega,\Lambda^{l+1} T\mathcal{M}))}.$$

Assuming that this is the case, we claim that δ_∂ factorizes as

$$
\frac{L^p_{\tan}(\partial\Omega,\Lambda^l T\mathcal{M})}{\delta_\partial\left(L^{p,\delta}_{\tan}(\partial\Omega,\Lambda^{l+1}T\mathcal{M})\right)} \xrightarrow{\ \Phi\ } \left(L^{q,0}_{\rm nor}(\partial\Omega,\Lambda^l T\mathcal{M})\right)^* \tag{9.6}
$$
$$
\xrightarrow{(d_\partial(\nu\wedge\cdot))^*} \left(\frac{\nu\vee L^{q,d}_{\rm nor}(\partial\Omega,\Lambda^l T\mathcal{M})}{\nu\vee L^{q,0}_{\rm nor}(\partial\Omega,\Lambda^l T\mathcal{M})}\right)^*,
$$

where Φ is the operator defined by

$$
\langle\!\langle \Phi([f]), g\rangle\!\rangle := -\int_{\partial\Omega}\langle \nu\wedge f, g\rangle\, d\sigma, \tag{9.7}
$$

for any

$$
[f]\in \frac{L^p_{\tan}(\partial\Omega,\Lambda^l T\mathcal{M})}{\delta_\partial\left(L^{p,\delta}_{\tan}(\partial\Omega,\Lambda^{l+1}T\mathcal{M})\right)},\qquad g\in L^{q,0}_{\rm nor}(\partial\Omega,\Lambda^l T\mathcal{M}).
$$

Note that Φ is well-defined since, by (9.1), the right side of (9.7) always vanishes when one pairs elements of the form $f=\delta_\partial h$, for $h\in L^{p,\delta}_{\tan}(\partial\Omega,\Lambda^{l+1}T\mathcal{M})$, with $g\in L^{q,0}_{\rm nor}(\partial\Omega,\Lambda^l T\mathcal{M})$. The linearity and continuity of this mapping are immediate.

Next, we claim that Φ is surjective. To see this, fix an arbitrary functional $\ell\in \left(L^{q,0}_{\rm nor}(\partial\Omega,\Lambda^l T\mathcal{M})\right)^*$. By Hahn-Banach's extension theorem and Riesz's representation theorem there exists $f\in L^p_{\rm nor}(\partial\Omega,\Lambda^{l+1}T\mathcal{M})$ which satisfies $\ell(h)=\int_{\partial\Omega}\langle f,h\rangle\,d\sigma$ for any $h\in L^{q,0}_{\rm nor}(\partial\Omega,\Lambda^{l+1}T\mathcal{M})$. It is then clear that Φ sends $[-\nu\vee f]$ into ℓ and, hence, Φ is onto.

We now examine the second horizontal arrow in sequence (9.6). By Proposition 9.3, the operator d_∂ in (9.5) is one-to-one and Fredholm. Since the operator

$$
\nu\wedge\cdot: \frac{\nu\vee L^{q,d}_{\rm nor}(\partial\Omega,\Lambda^l T\mathcal{M})}{\nu\vee L^{q,0}_{\rm nor}(\partial\Omega,\Lambda^l T\mathcal{M})}\longrightarrow \frac{L^{q,d}_{\rm nor}(\partial\Omega,\Lambda^l T\mathcal{M})}{L^{q,0}_{\rm nor}(\partial\Omega,\Lambda^l T\mathcal{M})}
$$

is an isomorphism, we deduce that $(d_\partial(\nu\wedge\cdot))^*$ in (9.6) is onto.

Combining all these facts, in order to show that δ_∂ in (9.4) is onto, we only need to prove that this operator factorizes as claimed in (9.6). To see this, for $f\in L^p_{\tan}(\partial\Omega,\Lambda^l T\mathcal{M})$ and $g\in L^{q,d}_{\rm nor}(\partial\Omega,\Lambda^l T\mathcal{M})$ we write

$$
\begin{aligned}
\langle\!\langle \delta_\partial([f]),[\nu\vee g]\rangle\!\rangle &= -\int_{\partial\Omega}\langle \nu\wedge f\,,\, d_\partial g\rangle\,d\sigma = \langle\!\langle \Phi([f])\,,\, d_\partial(\nu\wedge(\nu\vee g))\rangle\!\rangle \\
&= \langle\!\langle\, (d_\partial(\nu\wedge\cdot))^*\,\Phi([f])\,,\,[\nu\vee g]\rangle\!\rangle,
\end{aligned}\tag{9.8}
$$

i.e., $\delta_\partial=(d_\partial(\nu\wedge\cdot))^*\circ\Phi$, as desired. ∎

We next turn attention to the proof of Proposition 9.3. This involves certain algebraic topology tools, which we now begin to develop. The first task is to "localize" the definition of d_∂. More specifically, if $1<p<\infty$, for U an arbitrary, fixed open subset of $\partial\Omega$, we define

$$
\begin{aligned}
L^p_{\rm nor}(U,\Lambda^l T\mathcal{M},loc) := \{ f\in L^p(U,\Lambda^l T\mathcal{M},loc);\ \varphi f\in L^p_{\rm nor}(\partial\Omega,\Lambda^l T\mathcal{M})\\
\forall\,\varphi\in C^0(\mathcal{M},\mathbb{R}) \text{ such that } \operatorname{supp}\varphi\cap\partial\Omega\subseteq U\}.
\end{aligned}\tag{9.9}
$$

Let $1 \le l \le m-1$. For $f \in L^p_{\rm nor}(U, \Lambda^l T\mathcal{M}, loc)$ we say that $d_\partial f \in L^p_{\rm nor}(U, \Lambda^{l+1}\mathbb{R}, loc)$ if there exists $g \in L^p_{\rm nor}(U, \Lambda^{l+1}\mathbb{R}, loc)$ such that

$$\int_{\partial\Omega} \langle \delta\psi, f\rangle \, d\sigma = \int_{\partial\Omega} \langle \psi, g\rangle \, d\sigma$$

for any $\psi \in C^1(\mathcal{M}, \Lambda^{l+1}T\mathcal{M})$ such that $\operatorname{supp}\psi \cap \partial\Omega \subseteq U$. In this case, we set $d_\partial f := g$. We also agree that d_∂ maps m-forms into zero.

Next, for $1 \le l \le m$ and any open subset U of $\partial\Omega$, we define the additive Abelian group

$$\mathcal{L}^l_p(U) := \left\{ f \in L^p_{\rm nor}(U, \Lambda^l T\mathcal{M}, loc);\ d_\partial f \in L^p_{\rm nor}(U, \Lambda^{l+1}\mathbb{R}, loc) \right\}. \tag{9.10}$$

Since $\mathcal{L}^l_p(U)$ is a module over the algebra $\mathrm{Lip}(\partial\Omega)$, it follows that the family $\mathcal{L}^l_p := (\mathcal{L}^l_p(U))_U$, indexed by open subsets in $\partial\Omega$, is a *fine sheaf* on the topological space $\partial\Omega$.

Going further, for each U open subset of $\partial\Omega$ we let $d^l_\partial(U)$ map $\mathcal{L}^l_p(U)$ into $\mathcal{L}^{l+1}_p(U)$ by $d^l_\partial(U) f := d_\partial f$, for $l = 1, 2, \ldots$. Then $d^l_\partial := (d^l_\partial(U))_{U \text{ open in } \partial\Omega}$ is a (sheaf) morphism of $\mathcal{L}^l_p$ into $\mathcal{L}^{l+1}_p$ for each l. Since $d^{l+1}_\partial \circ d^l_\partial = 0$ for any l, this gives rise to the *complex* of sheaves

$$0 \longrightarrow \ker d^1_\partial \overset{\iota}{\longrightarrow} \mathcal{L}^1_p \overset{d^1_\partial}{\longrightarrow} \mathcal{L}^2_p \overset{d^2_\partial}{\longrightarrow} \mathcal{L}^3_p \overset{d^3_\partial}{\longrightarrow} \cdots \tag{9.11}$$

Let us denote by $\underline{\mathbb{R}}$ the sheaf of germs of locally constant functions on $\partial\Omega$.

LEMMA 9.4. *For any $1 < p < \infty$, the complex (9.11) provides a fine resolution of the sheaf $\underline{\mathbb{R}}\,\nu$ given by*

$$U \mapsto \{ f\nu;\ f \in \underline{\mathbb{R}}(U) \}. \tag{9.12}$$

The essential ingredient in the proof of this lemma is the *acyclicity* of the complex (9.11). The proof of this latter fact consists of reducing matters to proving an L^p-Poincaré type lemma for the operator $d_{\partial\Omega}$, the *intrinsic* exterior differential operator associated with the Lipschitz manifold $\partial\Omega$. In turn, this ultimately depends on clarifying the relationship between this latter operator and d_∂. In order not to disrupt the flow of the presentation at this point, we shall present the full argument in the Appendix B.

We shall use Lemma 9.4 in conjunction with the so-called abstract de Rham theorem, a version of which, well suited for our purposes, is recorded below.

THE ABSTRACT DE RHAM THEOREM. *Let $\mathcal{X}$ be a Hausdorff, paracompact topological space, and let $\mathcal{F}$ be a sheaf over $\mathcal{X}$. Also, let $\mathcal{L}^0, \mathcal{L}^1, ..$ be fine sheaves over $\mathcal{X}$ and, for $l = 0, 1, ...$, let $\vartheta_l : \mathcal{L}^l \to \mathcal{L}^{l+1}$ be sheaf homomorphisms such that the following is an exact complex:*

$$0 \longrightarrow \mathcal{F} \overset{\iota}{\hookrightarrow} \mathcal{L}^0 \overset{\vartheta_0}{\longrightarrow} \mathcal{L}^1 \overset{\vartheta_1}{\longrightarrow} \mathcal{L}^2 \overset{\vartheta_2}{\longrightarrow} \cdots \tag{9.13}$$

Then

$$H^l(\mathcal{X}; \mathcal{F}) \cong \frac{Ker(\vartheta_l : \mathcal{L}^l(\mathcal{X}) \longrightarrow \mathcal{L}^{l+1}(\mathcal{X}))}{Im(\vartheta_{l-1} : \mathcal{L}^{l-1}(\mathcal{X}) \longrightarrow \mathcal{L}^l(\mathcal{X}))}, \quad l = 1, 2, ...$$

See [Wa], Theorem 5.25, pp. 185 for a proof; cf. also [Go].

We are finally ready to present the

PROOF OF PROPOSITION 9.3. The abstract de Rham theorem applied to the boundary-complex (9.11) plus algebraic topology give the fundamental isomorphism

$$\frac{\operatorname{Ker}(d_{\partial}^{l+1}(\partial\Omega):\mathcal{L}_p^{l+1}(\partial\Omega)\to\mathcal{L}_p^{l+2}(\partial\Omega))}{\operatorname{Im}(d_{\partial}^{l}(\partial\Omega):\mathcal{L}_p^{l}(\partial\Omega)\to\mathcal{L}_p^{l+1}(\partial\Omega))}\cong H^l_{\rm sing}(\partial\Omega;\mathbb{R}) \tag{9.14}$$

valid for $l=1,2,..,m$ (recall that $H^l_{\rm sing}(\partial\Omega;\mathbb{R})$ is the l-th singular homology group of $\partial\Omega$ over $\mathbb{R}$, and that $b_l(\partial\Omega):=\dim H^l_{\rm sing}(\partial\Omega;\mathbb{R})$ is the l-th Betti number of $\partial\Omega$).

Clearly, since $\partial\Omega$ is compact, $\mathcal{L}_p^l(\partial\Omega)=L^{p,d}_{\rm nor}(\partial\Omega,\Lambda^lT\mathcal{M})$ for each l, whereas $d^l_\partial(\partial\Omega)$ is just the d_∂ operator introduced in connection with (5.14). Thus, the isomorphism above implies

$$\frac{L^{p,0}_{\rm nor}(\partial\Omega,\Lambda^{l+1}T\mathcal{M})}{d_\partial\Big(L^{p,d}_{\rm nor}(\partial\Omega,\Lambda^lT\mathcal{M})\Big)}\cong H^l_{\rm sing}(\partial\Omega;\mathbb{R}) \tag{9.15}$$

for $l=0,1,..,m$ and $1<p<\infty$.

Now, the closedness of the range of the operator d_∂ (in the context of Proposition 9.3) follows from (9.15), (5.6) and the fact that $\dim H^l_{\rm sing}(\partial\Omega;\mathbb{R})=b_l(\partial\Omega)<+\infty$. The remaining part in the statement of the theorem is immediate from (9.15) and the fact that the kernel of d_∂ acting on $L^{p,d}_{\rm nor}(\partial\Omega,\Lambda^lT\mathcal{M})$ is $L^{p,0}_{\rm nor}(\partial\Omega,\Lambda^lT\mathcal{M})$. This finishes the proof of Proposition 9.3. ■

We conclude this chapter by presenting a decomposition result of independent interest which is essentially a corollary of Proposition 9.1.

PROPOSITION 9.5. *For each $l\in\{0,1,...,m\}$, there holds*

$$\begin{aligned}L^2_{\rm tan}(\partial\Omega,\Lambda^lT\mathcal{M})=&\delta_\partial L^{2,\delta}_{\rm tan}(\partial\Omega,\Lambda^{l+1}T\mathcal{M})\oplus\big[\nu\vee d_\partial L^{2,d}_{\rm nor}(\partial\Omega,\Lambda^lT\mathcal{M})\big]\\&\oplus\big[L^{2,0}_{\rm tan}(\partial\Omega,\Lambda^lT\mathcal{M})\cap(\nu\vee L^{2,0}_{\rm nor}(\partial\Omega,\Lambda^{l+1}T\mathcal{M}))\big].\end{aligned} \tag{9.16}$$

A similar decomposition is valid for $L^2_{\rm nor}(\partial\Omega,\Lambda^lT\mathcal{M})$. In particular,

$$\begin{aligned}L^2(\partial\Omega,\Lambda^lT\mathcal{M})=&\delta_\partial L^{2,\delta}_{\rm tan}(\partial\Omega,\Lambda^{l+1}T\mathcal{M})\oplus\big[\nu\vee d_\partial L^{2,d}_{\rm nor}(\partial\Omega,\Lambda^lT\mathcal{M})\big]\\&\oplus d_\partial L^{2,d}_{\rm nor}(\partial\Omega,\Lambda^{l-1}T\mathcal{M})\oplus\big[\nu\wedge\delta_\partial L^{2,\delta}_{\rm tan}(\partial\Omega,\Lambda^lT\mathcal{M})\big]\\&\oplus\big[L^{2,0}_{\rm tan}(\partial\Omega,\Lambda^lT\mathcal{M})\cap(\nu\vee L^{2,0}_{\rm nor}(\partial\Omega,\Lambda^{l+1}T\mathcal{M}))\big]\\&\oplus\big[L^{2,0}_{\rm nor}(\partial\Omega,\Lambda^lT\mathcal{M})\cap(\nu\wedge L^{2,0}_{\rm tan}(\partial\Omega,\Lambda^{l-1}T\mathcal{M}))\big],\end{aligned} \tag{9.17}$$

where all direct sums are orthogonal.

PROOF. We shall only prove (9.16) as (9.17) is an immediate corollary of this. To this end, consider δ_∂ as a closed, densely defined, unbounded operator from $L^2_{\rm tan}(\partial\Omega,\Lambda^{l+1}T\mathcal{M})$ into $L^2_{\rm tan}(\partial\Omega,\Lambda^lT\mathcal{M})$ with domain $L^{2,\delta}_{\rm tan}(\partial\Omega,\Lambda^{l+1}T\mathcal{M})$. Clearly, its kernel is $L^{2,0}_{\rm tan}(\partial\Omega,\Lambda^{l+1}T\mathcal{M})$.

Now, by invoking Proposition 9.1, it is not difficult to check that δ_∂^*, the adjoint δ_∂ (in the sense of unbounded operators), has the domain $\nu\vee L^{2,d}_{\rm nor}(\partial\Omega,\Lambda^{l+1}T\mathcal{M})$ and that $\delta^*_\partial(\nu\vee g)=-\nu\vee d_\partial g$ for each $g\in L^{2,d}_{\rm nor}(\partial\Omega,\Lambda^{l+1}T\mathcal{M})$. In particular, $\operatorname{Ker}\delta^*_\partial=\nu\vee L^{2,0}_{\rm nor}(\partial\Omega,\Lambda^{l+1}T\mathcal{M})$.

At this point, relying on the fact that $\delta_\partial\circ\delta_\partial=0$ (in the sense of composition of unbounded operators) and plain functional analysis, it is trivial to check the version of (9.16) in which the first two summands are replaced by their respective L^2-closures. However, from (9.15) and the Hodge $*$-isomorphism, we see that $\delta_\partial L^{2,\delta}_{\tan}(\partial\Omega, \Lambda^{l+1}T\mathcal{M})$ has codimension $\dim H^{m-l-1}_{\text{sing}}(\partial\Omega;\mathbb{R}) < \infty$ in the space $L^{2,0}_{\tan}(\partial\Omega, \Lambda^l T\mathcal{M})$. Since the latter is a closed subspace of $L^2_{\tan}(\partial\Omega, \Lambda^l T\mathcal{M})$ it follows that δ_∂ has closed range. By Banach's closed range theorem, the same applies to δ^*_∂. This finishes the proof. ■

CHAPTER 10

Localization Arguments and the End of the Proof of Theorem 6.2

Throughout this chapter we continue to assume the usual set of hypotheses on $\mathcal{M}$ and the metric tensor, specified in Proposition 8.1. The main goal is to present the proof of the following (somewhat strengthened) version of Theorem 6.2:

THEOREM 10.1. *Let Ω be an arbitrary Lipschitz domain in the manifold $\mathcal{M}$, and let $V \in L^\infty(\mathcal{M})$ be a non-negative, scalar-valued potential that is > 0 in some open subset of $\mathcal{M}$. Then there exists $\varepsilon = \varepsilon(\Omega, V) > 0$ so that for $0 \le l \le m$ and $\lambda \in \mathbb{R}$, $|\lambda| \ge \frac{1}{2}$, the operators*

1. $\lambda I + M_l : L^p_{\rm tan}(\partial\Omega, \Lambda^l T\mathcal{M}) \to L^p_{\rm tan}(\partial\Omega, \Lambda^l T\mathcal{M})$,
2. $\lambda I + N_l : L^p_{\rm nor}(\partial\Omega, \Lambda^l T\mathcal{M}) \to L^p_{\rm nor}(\partial\Omega, \Lambda^l T\mathcal{M})$

are Fredholm with index zero on the indicated spaces whenever $2 - \varepsilon < p < 2 + \varepsilon$.

In passing, let us remark that this theorem immediately implies that $\lambda I + M_l\big(\nu \vee (\nu \wedge \cdot)\big)$ and $\lambda I + N_l\big(\nu \wedge (\nu \vee \cdot)\big)$ are Fredholm with index zero on the whole $L^p(\partial\Omega, \Lambda^l T\mathcal{M})$ whenever $2 - \varepsilon < p < 2 + \varepsilon$ and $\lambda \in \mathbb{R}$, $|\lambda| \ge \frac{1}{2}$.

PROOF. Since the claim is obviously true for $|\lambda|$ large and since

$$\big\{L^p_{\rm tan}(\partial\Omega, \Lambda^l T\mathcal{M})\big\}_{1<p<\infty}, \quad \text{and} \quad \big\{L^p_{\rm nor}(\partial\Omega, \Lambda^l T\mathcal{M})\big\}_{1<p<\infty}$$

are complex interpolation scales, known stability results of the Fredholm property and of the index (cf. [CS], [KM]) imply that it suffices to treat only the case $p = 2$, which we shall assume from now on. Also, by (6.12), it suffices to handle the case of the first operator only.

Continuing the series of reductions, we note that, by the comments in the first part of §8, we may restrict attention to the case when $1 < l < m$. In fact, since the operator in (1) is Fredholm with index zero for $l = 0$, by virtue of (6.12), the same can be said about the operator in (2) for $l = m$ and further, via (6.14), about $\lambda I + \tilde{M}_{m-1}$ on $L^2_{\rm tan}(\partial\Omega, \Lambda^l T\mathcal{M})$. Now, in view of (6.5), this latter operator is a compact perturbation of (1). The bottom line is that $l = m - 1$ is also covered and, hence, it is enough to consider only the case when $1 < l < m - 1$.

To proceed, we shall first assume that

$$(10.1) \qquad \mathcal{M} \text{ is a homology sphere.}$$

There are several consequences of major interest stemming from this supplementary hypothesis, which we now proceed to describe. First, the entire discussion in §6 can be carried out in the case when $V \equiv 0$ as well. This is because the Hodge Laplacian itself is now invertible on appropriate spaces of l-forms ($0 < l < m$), which, in turn, is a consequence of the absence of l-harmonic fields (by the classical de Rham-Hodge theorem asserting that the cohomology of a compact Riemannian manifold can be represented by harmonic forms). Second, corresponding to $V = 0$, we have several genuine commutation identities. Most notably,

$$(10.2)\qquad \begin{aligned} \delta_\partial M_l f &= M_{l-1}(\delta_\partial f) \text{ on } \partial\Omega, \ \forall\, f \in L^{2,\delta}_{\tan}(\partial\Omega, \Lambda^l T\mathcal{M}), \\ d_\partial N_l g &= N_{l+1}(d_\partial g) \text{ on } \partial\Omega, \ \forall\, g \in L^{2,d}_{\mathrm{nor}}(\partial\Omega, \Lambda^l T\mathcal{M}), \\ \delta \mathcal{S}_l f &= \mathcal{S}_l(\delta_\partial f) \text{ in } \Omega, \ \forall\, f \in L^{2,\delta}_{\tan}(\partial\Omega, \Lambda^l T\mathcal{M}), \\ d\mathcal{S}_l g &= \mathcal{S}_l(d_\partial g) \text{ in } \Omega, \ \forall\, g \in L^{2,d}_{\mathrm{nor}}(\partial\Omega, \Lambda^l T\mathcal{M}) \end{aligned}$$

(boundary versions of the last two identities are also valid). Also, from (6.13), the adjoint of $N_{l+1} : L^2_{\mathrm{nor}}(\partial\Omega, \Lambda^{l+1}T\mathcal{M}) \to L^2_{\mathrm{nor}}(\partial\Omega, \Lambda^{l+1}T\mathcal{M})$ is

$$(10.3)\qquad N^t_{l+1} g = \nu \wedge M_l(\nu \vee g), \quad g \in L^2_{\mathrm{nor}}(\partial\Omega, \Lambda^{l+1}T\mathcal{M}).$$

Another basic fact, seen from (10.2), is that $L^{2,0}_{\tan}(\partial\Omega, \Lambda^l T\mathcal{M})$ becomes an invariant subspace of M_l and we also claim that

$$(10.4)\qquad \lambda I + M_l : L^{2,0}_{\tan}(\partial\Omega, \Lambda^l T\mathcal{M}) \longrightarrow L^{2,0}_{\tan}(\partial\Omega, \Lambda^l T\mathcal{M})$$

is Fredholm with index zero, for any $\lambda \in \mathbb{R}$, $|\lambda| \geq \frac{1}{2}$. Indeed, this follows directly from (8.1) and the observation that, by (10.2), $\nu \vee \delta \mathcal{S}_l$ vanishes identically on $L^{2,0}_{\tan}(\partial\Omega, \Lambda^l T\mathcal{M})$. Consequently, combining this with (6.12), it follows that for any $\lambda \in \mathbb{R}$ with $|\lambda| \geq \frac{1}{2}$,

$$(10.5)\qquad \lambda I + N_l : L^{2,0}_{\mathrm{nor}}(\partial\Omega, \Lambda^l T\mathcal{M}) \longrightarrow L^{2,0}_{\mathrm{nor}}(\partial\Omega, \Lambda^l T\mathcal{M})$$

is also Fredholm with index zero.

The key idea is now to show that, in this topological context, the boundary derivative operator δ_∂, considered in the weak sense:

$$\delta_\partial : \frac{L^2_{\tan}(\partial\Omega, \Lambda^l T\mathcal{M})}{L^{2,0}_{\tan}(\partial\Omega, \Lambda^l T\mathcal{M})} \longrightarrow \left(\frac{\nu \vee L^{2,d}_{\mathrm{nor}}(\partial\Omega, \Lambda^l T\mathcal{M})}{\nu \vee L^{2,0}_{\mathrm{nor}}(\partial\Omega, \Lambda^l T\mathcal{M})} \right)^*$$

(cf. the discussion in Chapter 9) intertwines the operators

$$(10.6)\qquad \lambda I + M_l : \frac{L^2_{\tan}(\partial\Omega, \Lambda^l T\mathcal{M})}{L^{2,0}_{\tan}(\partial\Omega, \Lambda^l T\mathcal{M})} \longrightarrow \frac{L^2_{\tan}(\partial\Omega, \Lambda^l T\mathcal{M})}{L^{2,0}_{\tan}(\partial\Omega, \Lambda^l T\mathcal{M})}$$

and

(10.7)

$$(\nu \vee (\lambda I + N_l)\, \nu \wedge \cdot)^t : \left(\frac{\nu \vee L^{2,d}_{\mathrm{nor}}(\partial\Omega, \Lambda^l T\mathcal{M})}{\nu \vee L^{2,0}_{\mathrm{nor}}(\partial\Omega, \Lambda^l T\mathcal{M})} \right)^* \to \left(\frac{\nu \vee L^{2,d}_{\mathrm{nor}}(\partial\Omega, \Lambda^l T\mathcal{M})}{\nu \vee L^{2,0}_{\mathrm{nor}}(\partial\Omega, \Lambda^l T\mathcal{M})} \right)^*.$$

That is,

$$(10.8)\qquad \delta_\partial(\lambda I + M_l) = (\nu \vee (\lambda I + N_l)\, \nu \wedge \cdot)^t\, \delta_\partial.$$

Here the superscripts $*$, t indicate, respectively, the dual space and the adjoint operator. Also, the operators above act on quotient spaces in a natural way (i.e., they commute with the projection operators on the quotient spaces). Parenthetically, let us point out that an L^p-version of this result is also valid; in particular, this yields the spectral identity

$$\sigma\left(M_l; \frac{L^p_{\tan}(\partial\Omega, \Lambda^\ell T\mathcal{M})}{L^{p,0}_{\tan}(\partial\Omega, \Lambda^\ell T\mathcal{M})} \right) = \sigma\left(N_l; \frac{L^{q,d}_{\mathrm{nor}}(\partial\Omega, \Lambda^\ell T\mathcal{M})}{L^{q,0}_{\mathrm{nor}}(\partial\Omega, \Lambda^\ell T\mathcal{M})} \right)^c,$$

where $1/p + 1/q = 1$ and the superscript c stands for complex conjugation.

Turning now to the task of checking the intertwining property alluded to before, we let $[\cdot]$ stand for the projection operator onto the various quotient spaces above. Also, fix

$$[f] \in \frac{L^2_{\tan}(\partial\Omega, \Lambda^l T\mathcal{M})}{L^{2,0}_{\tan}(\partial\Omega, \Lambda^l T\mathcal{M})},$$

and pick an arbitrary

$$[\nu \vee g] \in \frac{\nu \vee L^{2,d}_{\rm nor}(\partial\Omega, \Lambda^l T\mathcal{M})}{\nu \vee L^{2,0}_{\rm nor}(\partial\Omega, \Lambda^l T\mathcal{M})}.$$

Then, if $\langle\!\langle \cdot, \cdot \rangle\!\rangle$ denotes the natural pairing between various dual spaces, we have

$$\begin{aligned}
&\langle\!\langle (\nu \vee (\lambda I + N_l)\, \nu \wedge \cdot)^t (\delta_\partial [f]) \,,\, [\nu \vee g] \rangle\!\rangle \\
&\quad = \langle\!\langle \delta_\partial f \,,\, \nu \vee (\lambda I + N_l)(\nu \wedge \cdot [\nu \vee g]) \rangle\!\rangle \\
&\quad = \langle\!\langle \delta_\partial f \,,\, [\nu \vee (\lambda I + N_l) g] \rangle\!\rangle \\
&\quad = -\int_{\partial\Omega} \langle \nu \wedge f \,,\, d_\partial (\lambda I + N_l) g \rangle \, d\sigma && \text{(by (9.3))} \\
&\quad = -\int_{\partial\Omega} \langle \nu \wedge f \,,\, (\lambda I + N_{l+1})(d_\partial g) \rangle \, d\sigma && \text{(by (10.2))} \\
&\quad = -\int_{\partial\Omega} \langle (\lambda I + N_{l+1})^t \, (\nu \wedge f) \,,\, d_\partial g \rangle \, d\sigma \\
&\quad = -\int_{\partial\Omega} \langle \nu \wedge (\lambda I + M_l)(\nu \vee (\nu \wedge f)) \,,\, d_\partial g \rangle \, d\sigma && \text{(by (10.3))} \\
&\quad = -\int_{\partial\Omega} \langle \nu \wedge (\lambda I + M_l) f \,,\, d_\partial g \rangle \, d\sigma \\
&\quad = \langle\!\langle \delta_\partial [(\lambda I + M_l) f] \,,\, [\nu \vee g] \rangle\!\rangle && \text{(by (9.3)).}
\end{aligned}$$

Hence, $\delta_\partial (\lambda I + M_l)[f] = (\nu \vee (\lambda I + N_l)\nu \wedge \cdot)^t (\delta_\partial [f])$ which proves (10.8).

The basic feature of the intertwining identity (10.8) is that it enables us to relate the action of the operator $\lambda I + M_l$ on (a quotient space involving) the "large" space $L^2_{\tan}(\partial\Omega, \Lambda^l T\mathcal{M})$ to the action of the operator $\lambda I + N_l$ on (quotient spaces involving) the "regularity" spaces $L^{2,\delta}_{\tan}(\partial\Omega, \Lambda^l T\mathcal{M})$ and $L^{2,0}_{\tan}(\partial\Omega, \Lambda^l T\mathcal{M})$. The important point is that we already know from Theorem 8.2 and elementary functional analysis that the latter operator, i.e., (10.7), is Fredholm with index zero on $L^{2,d}_{\rm nor}(\partial\Omega, \Lambda^l T\mathcal{M})$ as well as on $L^{2,0}_{\rm nor}(\partial\Omega, \Lambda^l T\mathcal{M})$. Thus, by virtue of Theorem 9.2, it follows that the operator in (10.6) is also Fredholm with index zero.

Restoring the original $L^2_{\tan}(\partial\Omega, \Lambda^l T\mathcal{M})$ space is done by invoking a functional analytic argument which we present in abstract form in the following lemma (whose proof is an exercise).

LEMMA 10.2. *Let $\mathcal{X}$, $\mathcal{Y}$, $\mathcal{Z}$ be Banach spaces and consider the commutative diagram*

$$\begin{array}{ccccccccc}
0 & \longrightarrow & \mathcal{X} & \longrightarrow & \mathcal{Y} & \longrightarrow & \mathcal{Z} & \longrightarrow & 0 \\
 & & \downarrow & & \downarrow & & \downarrow & & \\
0 & \longrightarrow & \mathcal{X} & \longrightarrow & \mathcal{Y} & \longrightarrow & \mathcal{Z} & \longrightarrow & 0
\end{array}$$

where all arrows are linear, bounded and the horizontal sequences are exact. Then, if two vertical arrows are Fredholm operators then so is the third one and the index of the middle arrow is the sum of the indexes of the other two vertical arrows.

Of course, in our case we apply this lemma taking the first two horizontal arrows to be inclusions and the next two to be projections (in each short sequence),

$$\mathcal{X} := L^{2,0}_{\tan}(\partial\Omega, \Lambda^l T\mathcal{M}), \quad \mathcal{Y} := L^2_{\tan}(\partial\Omega, \Lambda^l T\mathcal{M}), \quad \mathcal{Z} := \frac{L^2_{\tan}(\partial\Omega, \Lambda^l T\mathcal{M})}{L^{2,0}_{\tan}(\partial\Omega, \Lambda^l T\mathcal{M})},$$

while all vertical arrows are taken to be natural manifestations of the operator $\lambda I + M_l$ on the spaces listed above. Thus, at this point, we have proved that

$$\text{(10.9)} \qquad \begin{array}{l} (10.1) \Longrightarrow \lambda I + M_l \text{ is Fredholm with index zero} \\ \qquad\qquad \text{on } L^2_{\tan}(\partial\Omega, \Lambda^l T\mathcal{M}), \ \forall\, |\lambda| \geq \frac{1}{2}. \end{array}$$

Next we indicate how dispense with the extra assumption (10.1) in the implication above. To this end, we consider an open, finite covering $\{U_j\}_{1\leq j\leq J}$ of $\partial\Omega$ with domains of coordinate charts in $\mathcal{M}$, all of which are homeomorphic with the unit ball in $\mathbb{R}^m$. The fundamental fact we shall use here is that each U_j can be embedded isometrically in a compact, boundaryless Riemannian manifold $\mathcal{M}_j$ having the same dimension m (e.g., the so-called *double* obtained by taking two replicas of U_j with opposite orientations and "gluing" them together by identifying boundary points). Select also, for each j, a Lipschitz domain Ω_j in $\mathcal{M}_j$ such that $\{\partial\Omega_j \cap \partial\Omega\}_j$ makes up an open cover of $\partial\Omega$. Because of our assumptions, each $\mathcal{M}_j$ satisfies (10.1) so that, if we denote by $M_{l,j}$ the operator constructed as in (6.9) but for the domain $\partial\Omega_j$ in the manifold $\mathcal{M}_j$ and with $V \equiv 0$, then $\lambda I + M_{l,j}$ is Fredholm with index zero on $L^2_{\tan}(\partial\Omega_j, \Lambda^l T\mathcal{M}_j)$ for all real λ's with $|\lambda| \geq \frac{1}{2}$. This implies that, for each such λ, there exists $C = C_\lambda > 0$ such that

$$\text{(10.10)} \qquad \|f\|_{L^2(\partial\Omega_j,\Lambda^l T\mathcal{M}_j)} \leq C\|(\lambda I + M_{l,j})f\|_{L^2(\partial\Omega_j,\Lambda^l T\mathcal{M}_j)} + \|\mathrm{Comp}(f)\|$$

uniformly for $f \in L^2_{\tan}(\partial\Omega_j, \Lambda^l T\mathcal{M}_j)$ where, hereafter, Comp denotes generic compact operators between Banach spaces. Let $\{\xi_j\}_{1\leq j\leq J}$ be Lipschitz continuous functions on $\mathcal{M}$ which form a partition of unity subordinated to $\{U_j\}_{1\leq j\leq J}$ and such that $\operatorname{supp}\xi_j \cap \partial\Omega \subseteq \partial\Omega_j$ for each j. Also, for each j, let η_j be a Lipschitz function on $\mathcal{M}$ supported in U_j and such that $\eta_j \equiv 1$ in a neighborhood of $\operatorname{supp}\xi_j$. Finally, fix $\lambda \in \mathbb{R} \setminus (-\frac{1}{2}, \frac{1}{2})$ and let $f \in L^2_{\tan}(\partial\Omega, \Lambda^l T\mathcal{M})$ be arbitrary. We have

$$\text{(10.11)} \qquad \begin{aligned} &\|f\|_{L^2(\partial\Omega,\Lambda^l T\mathcal{M})} \\ &\leq \sum_{j=1}^J \|\xi_j f\|_{L^2(\partial\Omega,\Lambda^l T\mathcal{M})} = \sum_{j=1}^J \|\xi_j f\|_{L^2(\partial\Omega_j,\Lambda^l T\mathcal{M}_j)} \\ &\leq C\sum_{j=1}^J \|(\lambda I + M_{l,j})(\xi_j f)\|_{L^2(\partial\Omega_j,\Lambda^l T\mathcal{M}_j)} + \|\mathrm{Comp}\,(f)\| \\ &\leq C\sum_{j=1}^J \|\eta_j(\lambda I + M_{l,j})(\xi_j f)\|_{L^2(\partial\Omega_j,\Lambda^l T\mathcal{M}_j)} \\ &\quad + C\sum_{j=1}^J \|(1-\eta_j)(\lambda I + M_{l,j})(\xi_j f)\|_{L^2(\partial\Omega_j,\Lambda^l T\mathcal{M}_j)} + \|\mathrm{Comp}\,(f)\|. \end{aligned}$$

The following remarks are relevant for continuing this calculation. In the first place, for each j, the operator $(1-\eta_j)(\lambda I + M_{l,j})\xi_j$ has a bounded kernel and, hence, is compact from $L^2(\partial\Omega, \Lambda^l T\mathcal{M})$ into $L^2(\partial\Omega_j, \Lambda^l T\mathcal{M}_j)$. In the second place, from (6.1) and (3.18), the difference

$$\eta_j(\lambda I + M_{l,j})\xi_j - \eta_j(\lambda I + M_l)\xi_j \tag{10.12}$$

is a compact operator from $L^2(\partial\Omega, \Lambda^l T\mathcal{M})$ into $L^2(\partial\Omega \cap \partial\Omega_j, \Lambda^l T\mathcal{M})$. Thirdly, the commutator $[M_l, \xi_j]$ has a weakly singular kernel, hence defines a compact operator on $L^2(\partial\Omega, \Lambda^l T\mathcal{M})$. With these facts in mind, returning to (10.11), we have

$$\begin{aligned} \|f\|_{L^2(\partial\Omega,\Lambda^l T\mathcal{M})} &\leq C\sum_{j=1}^{J} \|\eta_j(\lambda I + M_l)(\xi_j f)\|_{L^2(\partial\Omega,\Lambda^l T\mathcal{M})} + \|\mathrm{Comp}\,(f)\| \\ &\leq C\sum_{j=1}^{J} \|\xi_j(\lambda I + M_l) f\|_{L^2(\partial\Omega,\Lambda^l T\mathcal{M})} + \|\mathrm{Comp}\,(f)\| \\ &\leq C\|(\lambda I + M_l) f\|_{L^2(\partial\Omega,\Lambda^l T\mathcal{M})} + \|\mathrm{Comp}\,(f)\|. \end{aligned} \tag{10.13}$$

Thus, $\lambda I + M_l$ is bounded from below modulo compact operators on $L^2(\partial\Omega, \Lambda^l T\mathcal{M})$ for each $\lambda \in \mathbb{R}\setminus(-\frac{1}{2}, \frac{1}{2})$. As before, functional analysis implies that these operators (indexed by λ) are Fredholm with index zero, as desired.

Finally, this readily implies the same conclusion for the operator $\lambda I + N_l$, mapping the space $L^2_{\rm nor}(\partial\Omega, \Lambda^l T\mathcal{M})$ into itself (by a simple application of the Hodge $*$-isomorphism; cf. (6.12)). The proof of the theorem is therefore complete. ∎

We conclude this chapter with a few corollaries that are going to be of importance for us. The first regards the operators $\tilde{M}_l$ and $\tilde{N}_l$ introduced in (6.10).

COROLLARY 10.3. *With the same hypotheses as in Theorem 10.1, for any* $\lambda \in \mathbb{R}$ *with* $|\lambda| \geq \frac{1}{2}$ *the operators* $\lambda I + \tilde{M}_l$ *and* $\lambda I + \tilde{N}_l$ *are Fredholm with index zero when acting on* $L^p_{\rm tan}(\partial\Omega, \Lambda^l T\mathcal{M})$ *and* $L^p_{\rm nor}(\partial\Omega, \Lambda^l T\mathcal{M})$*, respectively, where* $\varepsilon > 0$ *is as in the conclusion of Theorem 10.1.*

PROOF. This is seen from (6.1), (6.3), (6.5) and Theorem 10.1. ∎

The second one augments and extends the results presented in Corollary 8.3.

COROLLARY 10.4. *Suppose that the potential* V *is a strictly positive constant and let* $\varepsilon = \varepsilon(\Omega, V) > 0$ *be as in the conclusion of Theorem 10.1. Then, for each* l *and any* $\lambda \in \mathbb{R}$ *with* $|\lambda| \geq \frac{1}{2}$*, the operators*

$$\begin{aligned} \lambda I + M_l &: L^p_{\rm tan}(\partial\Omega, \Lambda^l T\mathcal{M}) \to L^p_{\rm tan}(\partial\Omega, \Lambda^l T\mathcal{M}), \\ \lambda I + N_l &: L^p_{\rm nor}(\partial\Omega, \Lambda^l T\mathcal{M}) \to L^p_{\rm nor}(\partial\Omega, \Lambda^l T\mathcal{M}) \end{aligned} \tag{10.14}$$

and

$$\begin{aligned} \lambda I + M_l &: L^{p,\delta}_{\rm tan}(\partial\Omega, \Lambda^l T\mathcal{M}) \to L^{p,\delta}_{\rm tan}(\partial\Omega, \Lambda^l T\mathcal{M}), \\ \lambda I + N_l &: L^{p,d}_{\rm nor}(\partial\Omega, \Lambda^l T\mathcal{M}) \to L^{p,d}_{\rm nor}(\partial\Omega, \Lambda^l T\mathcal{M}) \end{aligned} \tag{10.15}$$

are invertible for each $2 - \varepsilon < p < 2 + \varepsilon$*.*

PROOF. First consider the operators in (10.14). Since $L^{2,\delta}_{\tan}(\partial\Omega,\Lambda^l T\mathcal{M})$ and $L^{2,d}_{\rm nor}(\partial\Omega,\Lambda^l T\mathcal{M})$ can be seen to embed densely into the spaces $L^p_{\tan}(\partial\Omega,\Lambda^l T\mathcal{M})$ and $L^p_{\rm nor}(\partial\Omega,\Lambda^l T\mathcal{M})$, respectively, the case $p=2$ follows from Corollary 8.3 and Theorem 10.1. With this at hand, the first part of the corollary follows easily with the aid of Theorem 10.1 since, depending on whether $p<2$ or $p>2$, the operators in (10.14) are with dense range or one-to-one, respectively.

Turning to the operators in (10.15) we first note that, from what we have proved so far,

$$\|f\|_{L^p(\partial\Omega,\Lambda^l T\mathcal{M})} \leq C(p,\lambda)\|(\lambda I+M_l)f\|_{L^p(\partial\Omega,\Lambda^l T\mathcal{M})} \tag{10.16}$$

for any $l=0,1,...,m$, $2-\varepsilon<p<2+\varepsilon$, $\lambda\in\mathbb{R}$ with $|\lambda|\geq\frac{1}{2}$, uniformly for $f\in L^p_{\tan}(\partial\Omega,\Lambda^l T\mathcal{M})$. Now, if f is actually in $L^{p,\delta}_{\tan}(\partial\Omega,\Lambda^l T\mathcal{M})$, we may also write the $(l-1)$-version of the estimate (10.16) for $\delta_\partial f\in L^p_{\tan}(\partial\Omega,\Lambda^{l-1}T\mathcal{M})$. Adding the two estimates and invoking the identity (6.19), we obtain that

$$\|f\|_{L^{p,\delta}_{\tan}(\partial\Omega,\Lambda^l T\mathcal{M})} \leq C(p,\lambda)\|(\lambda I+M_l)f\|_{L^{p,\delta}_{\tan}(\partial\Omega,\Lambda^l T\mathcal{M})} + \|\mathrm{Comp}\,(f)\| \tag{10.17}$$

if $2-\varepsilon<p<2+\varepsilon$ and $\lambda\in\mathbb{R}$ with $|\lambda|\geq\frac{1}{2}$, uniformly for $f\in L^{p,\delta}_{\tan}(\partial\Omega,\Lambda^l T\mathcal{M})$; here Comp denotes generic compact operators on L^p. Thus, the operators in (10.15) are Fredholm with index zero (a result which is in fact independent of the nature of the potential V) as long as $2-\varepsilon<p<2+\varepsilon$. Now, the proof can be completed as before, on account of Corollary 8.3. ■

Finally, we present a regularity result.

COROLLARY 10.5. *Once again, assume the same hypotheses and notation as in Corollary 10.4. Also, fix $2-\varepsilon<p\leq q<2+\varepsilon$. Then if $f\in L^p_{\tan}(\partial\Omega,\Lambda^l T\mathcal{M})$ is such that $(\lambda I+M_l)f\in L^{q,\delta}_{\tan}(\partial\Omega,\Lambda^l T\mathcal{M})$ then, necessarily, $f\in L^{q,\delta}_{\tan}(\partial\Omega,\Lambda^l T\mathcal{M})$.*

Furthermore, a similar result is valid for the operators $\lambda I+\tilde{M}_l$, $\lambda I+N_l$ and $\lambda I+\tilde{N}_l$.

PROOF. Suppose, for instance, that $f\in L^p_{\tan}(\partial\Omega,\Lambda^l T\mathcal{M})$ is such that $(\lambda I+M_l)f=:g\in L^{q,\delta}_{\tan}(\partial\Omega,\Lambda^l T\mathcal{M})$. Also, let M^0_l be the operator M_l corresponding to a constant, strictly positive potential V. The key idea is that, by (6.1), (6.15) and (3.18), $M^0_l-M_l$ is a "smoothing" operator, mapping $L^p_{\tan}(\partial\Omega,\Lambda^l T\mathcal{M})$ boundedly into $L^{q,\delta}_{\tan}(\partial\Omega,\Lambda^l T\mathcal{M})$ for $|p-q|$ sufficiently small. Thus, if $2-\varepsilon<p\leq q<2+\varepsilon$, $\varepsilon>0$ small, Corollary 10.4 gives

$$f=(\lambda I+M^0_l)^{-1}(g+M^0_l f-M_l f)\in L^{q,\delta}_{\tan}(\partial\Omega,\Lambda^l T\mathcal{M}),$$

as desired. ■

REMARKS. (i). In the case when $\partial\Omega\in C^1$, the methods in [FJR] and [JM] can be used to show that the results in this chapter continue to hold for the full range $1<p<\infty$.

(ii). An interesting conjecture is whether the operators $\pm\frac{1}{2}I+M_l$ are Fredholm with index zero on the spaces of "regular" forms $L^{p,\delta}_{\tan}(\partial\Omega,\Lambda^l T\mathcal{M})$ for each $1<p\leq 2$, $l\in\{0,1,...,m\}$ and for any Lipschitz domain $\Omega\subset\mathcal{M}$. At least in the flat Euclidean setting, this is true when $l=0$ ([DK]), $l=1$ ([MMP]), $l=m-1$ ([DK])

and $l = m$ (trivially). Let us also point out that, as has been shown in [MMP], this conjecture is false at the level of $L^p_{\tan}(\partial\Omega, \Lambda^1 T\mathcal{M})$ with $p \neq 2$.

CHAPTER 11

Harmonic Fields on Lipschitz Domains

We continue to maintain the same hypotheses on $\mathcal{M}$ and Ω as in the previous chapters. Here, the focus is the space of l-harmonic fields with vanishing normal component on $\partial\Omega$. Recall, from (5.7), that this is

$$\mathcal{H}_\vee^{l;p}(\Omega) := \{u \in C^1(\Omega, \Lambda^l T\mathcal{M});\ \mathcal{N}(u) \in L^p(\partial\Omega),$$
$$du = 0,\ \delta u = 0 \text{ in } \Omega,\ \nu \vee u\big|_{\partial\Omega} = 0\}.$$

We aim at proving the existence of a unique such harmonic field with arbitrarily preassigned periods. This is made precise in the following theorem.

THEOREM 11.1. *Fix an arbitrary* $l \in \{0, 1, ..., m\}$ *and let* $\{\gamma_j\}_{j=1,...,b_l(\Omega)}$ *be smooth* l*-cycles in* Ω *such that* $\{[\gamma_j]\}_{j=1,...,b_l(\Omega)}$ *is a basis for* $H^l_{\rm sing}(\Omega;\mathbb{R})$*, the* l*-th singular homology group of* Ω *over the reals. Also, denote by* ι *the inclusion of each* γ_j *into* $\mathcal{M}$*.*

Then there exists $\varepsilon = \varepsilon(\Omega) > 0$ *so that* $\mathcal{H}_\vee^{l;p}(\Omega)$ *is independent of* $p \in (2 - \varepsilon, 2 + \varepsilon)$*. (For simplicity, we shall occasionally drop the superscript* p *assuming, implicitly, that* $|2 - p| < \varepsilon$*.)*

Also, with p *as above, the mapping*

$$\Phi_l : \mathcal{H}_\vee^l(\Omega) \to \mathbb{R}^{b_l(\Omega)} \tag{11.1}$$

assigning to each harmonic field its periods, i.e.,

$$\Phi_l(u) := \left(\int_{\gamma_j} \iota^* u\right)_{j=1,2,...,b_l(\Omega)}, \tag{11.2}$$

is an isomorphism for each l*. In particular,*

$$\dim \mathcal{H}_\vee^l(\Omega) = b_l(\Omega), \quad \text{for } l = 0, 1, ..., m. \tag{11.3}$$

Note that (as is well known) the definition of Φ_l is independent of the choice of the basis $\{[\gamma_j]\}_j$ in $H^l_{\rm sing}(\Omega;\mathbb{R})$. Also, by (11.3) and the Hodge $*$-isomorphism, we have for $\mathcal{H}_\wedge^l(\Omega)$, given by (5.13), the result

$$\dim \mathcal{H}_\wedge^l(\Omega) = b_{m-l}(\Omega), \quad \text{for } l = 0, 1, ..., m.$$

The proof of this theorem rests on three major ingredients:

1. A *regularity result*, to the effect that if

$$\mathbb{H}_\vee^{l;p}(\Omega) := \left\{u \in L^p(\Omega, \Lambda^l T\mathcal{M});\ du = 0,\ \delta u = 0 \text{ in } \Omega,\ \nu \vee u = 0 \text{ on } \partial\Omega\right\}, \tag{11.4}$$

then

$$\mathbb{H}_\vee^{l;p}(\Omega) = \mathcal{H}_\vee^{l;p}(\Omega) = \mathcal{H}_\vee^{l;2}(\Omega) \text{ for } l = 0, 1, ..., m \text{ and } 2 - \varepsilon < p < 2 + \varepsilon. \tag{11.5}$$

Here $\nu \vee u = 0$ on $\partial\Omega$ is to be interpreted in a weak, variational sense explained in (11.8) below.

2. A *Hodge type decomposition* on arbitrary Lipschitz domains, guaranteeing that

$$\text{Ker}\, d_{l,2} = \text{Im}\, d_{l-1,2} \oplus \mathbb{H}_{\vee}^{l,2}(\Omega), \tag{11.6}$$

where $d_{l,p} : L^p(\Omega, \Lambda^l T\mathcal{M}) \to L^p(\Omega, \Lambda^{l+1} T\mathcal{M})$, $1 < p < \infty$, is the realization of the exterior derivative operator d as a maximally closed unbounded operator (with natural domain).

3. The classical *de Rham theorem* ([DR1], cf. also Chapter II of [Hod2], pp. 92–100), asserting that on a smooth, compact, boundaryless manifold $\mathcal{M}$ there exists a smooth, closed l-form having a priori prescribed periods on $b_l(\mathcal{M})$ independent l-cycles of $\mathcal{M}$.

Accepting these results for the time being, we shall now proceed to present the

PROOF OF THEOREM 11.1. First we shall deal with (11.3) which is a direct consequence of (11.5)–(11.6) and the ($p = 2$ case of the) isomorphism

$$\frac{\{u \in L^p(\Omega, \Lambda^l T\mathcal{M});\, du = 0 \text{ in } \Omega\}}{d\{v \in L^p(\Omega, \Lambda^{l-1} T\mathcal{M});\, dv \in L^p(\Omega, \Lambda^l T\mathcal{M})\}} \cong H_{\text{sing}}^l(\Omega; \mathbb{R}), \quad 1 < p < \infty. \tag{11.7}$$

In turn, this is a direct consequence of the abstract de Rham theorem (cf. §9) applied to the fine sheaves $\mathcal{L}_p^l = \big(\mathcal{L}_p^l(U)\big)_U$, indexed by open subsets in (the relative topology of) $\overline{\Omega}$, defined by

$$\mathcal{L}_p^l(U) := \{u \in L_{\text{loc}}^p(U, \Lambda^l T\mathcal{M});\, du \in L_{\text{loc}}^p(U, \Lambda^{l+1} T\mathcal{M})\}$$

and the sheaf morphisms $d : \mathcal{L}_p^l \to \mathcal{L}_p^{l+1}$, where d_U is simply the restriction of d to the interior of U in $\mathcal{M}$. Indeed, the only thing to be checked is the exactness on stalks. However, the L^p-Poincaré type lemma, $1 < p < \infty$, to be verified in this context is clearly invariant under pull-back via bi-Lipschitz homeomorphisms and, hence, can be transported to a ball in $\mathbb{R}^m$. In the latter case, such a result follows easily, e.g., via standard Hodge theory (alternatively, one can use Hodge decompositions in smooth domains, or the standard proof of the Poincaré lemma can be adapted to this setting by using a mollifying argument). This completes the proof of (11.7). Note that this proof utilizes only the Lipschitz structure of the underlying Riemannian manifold (an alternative proof can be given starting from the observation that the left side of (11.7) is independent of the metric and then working in the smooth context). As indicated before, (11.3) also follows.

Turning attention to the mapping (11.1)–(11.2) (and considering that $p = 2$), the idea is to prove that Φ_l is onto. Of course, by (11.3) and linearity, this suffices to conclude that this mapping is in fact an isomorphism. To see this, fix some arbitrary real numbers $\beta_1, ..., \beta_{b_l(\Omega)}$. We shall use de Rham's theorem mentioned a few paragraphs above. To achieve a setting in which this applies, we first embed Ω in a smooth domain $U \subseteq \mathcal{M}$ and then take $\mathcal{U}$ to be the double of U. A basic feature is that cycles which are independent with respect to homology in U remain so in $\mathcal{U}$. Hence, since matters can be arranged so that $\{[\gamma_j]\}_{j=1}^{b_l(\Omega)}$ still form a basis for $H_{\text{sing}}^l(U; \mathbb{R})$ it follows that $\{[\gamma_j]\}_{j=1}^{b_l(\Omega)}$ are linearly independent in $H_{\text{sing}}^l(\mathcal{U}; \mathbb{R})$. We complete them to a basis in $H_{\text{sing}}^l(\mathcal{U}; \mathbb{R})$ and apply the aforementioned theorem of de Rham for the periods $\alpha_j := \beta_j$ when $j = 1, ..., b_l(\Omega)$ and $\alpha_j := 0$ when $j = b_l(\Omega) + 1, ..., b_l(\mathcal{U})$. This yields a smooth, closed l-form w on $\mathcal{U}$ such that $\int_{\gamma_j} \iota^* w = \alpha_j$ for $j = 1, ..., b_l(\mathcal{U})$. Further, if we decompose $w|_\Omega \in C^1(\overline{\Omega}, \Lambda^l T\mathcal{M})$ as $w|_\Omega = v \oplus u$ according to (11.6) then, by Stokes's theorem, $\Phi_l(u) = (\beta_j)_{j=1,...,b_l(\Omega)}$. Thus Φ_l is onto and the proof of Theorem 11.1 is finished. ∎

In the second part of this chapter we present the proofs of the first two auxiliary results used in the proof of Theorem 11.1. First, we discuss a regularity theorem from which (11.5) will follow easily. Before we state it, we make a couple of definitions. Let Ω be a Lipschitz subdomain of $\mathcal{M}$ and denote the outgoing unit conormal on $\partial\Omega$ by ν. Also, recall the Besov scale of spaces $B^p_s(\partial\Omega)$ on $\partial\Omega$. Consider now $1 \leq l \leq m$ and $u \in L^p(\Omega, \Lambda^l T\mathcal{M})$ such that $\delta u \in L^p(\Omega, \Lambda^{l-1} T\mathcal{M})$ for some $1 < p < \infty$. Then we define the distribution $\nu \vee u$ on $\partial\Omega$ by requiring that

$$\langle\langle \nu \vee u, \varphi \rangle\rangle := -\iint_\Omega \langle \delta u, v\rangle \, d\text{Vol} + \iint_\Omega \langle u, dv\rangle \, d\text{Vol}, \tag{11.8}$$

for any $v \in H^{1,q}(\Omega, \Lambda^{l-1} T\mathcal{M})$, $1/p + 1/q = 1$, with $\mathrm{Tr}\, v = \varphi$. Here, $\mathrm{Tr} : C^0(\bar{\Omega}, \Lambda^l T\mathcal{M}) \to C^0(\partial\Omega, \Lambda^l T\mathcal{M})$ is the ordinary trace operator, which further extends from $H^{s,p}(\Omega, \Lambda^l T\mathcal{M})$ into $B^p_{s-1/p}(\partial\Omega, \Lambda^l T\mathcal{M})$ for any $s \in (\frac{1}{p}, 1 + \frac{1}{p})$, $1 < p < \infty$.

Thus, the right side of (11.8) is well defined for $\varphi \in B^q_{1/p}(\partial\Omega, \Lambda^{l-1} T\mathcal{M})$, independently of the choice of such v, so we have

$$\nu \vee u \in B^p_{-1/p}(\partial\Omega, \Lambda^{l-1} T\mathcal{M}) \tag{11.9}$$

with naturally accompanying estimates. Further, if $u \in L^p(\Omega, \Lambda^l T\mathcal{M})$ is such that $du \in L^p(\Omega, \Lambda^{l-1} T\mathcal{M})$ then we can define the distribution $\nu \wedge u$ by a similar procedure or, equivalently, set $\nu \wedge u := (-1)^{1+m(l+1)} * (\nu \vee *u)$. Once again,

$$\nu \wedge u \in B^p_{-1/p}(\partial\Omega, \Lambda^{l+1} T\mathcal{M}) \tag{11.10}$$

plus natural estimates.

Finally, for each $0 \leq l \leq m$ and each V as in the first part of § 6, we let Π_l stand for the volume potential operator mapping l-forms into l-forms formally by

$$\Pi_l u(x) := \iint_\Omega \langle \Gamma_l(x, y), u(y)\rangle \, d\text{Vol}(y), \qquad x \in \Omega, \tag{11.11}$$

where, as in §6, Γ_l is the Schwartz kernel of $(\Delta - V)^{-1}$. Recall, from Theorem 2.9, that

$$\Pi_l : L^p(\Omega, \Lambda^l T\mathcal{M}) \to H^{2,p}(\Omega, \Lambda^l T\mathcal{M}) \tag{11.12}$$

is a bounded operator for any $1 < p < r$, where $r > m$ is as in (4.10).

THEOREM 11.2. *For any Ω arbitrary Lipschitz domain in $\mathcal{M}$ there exists $\varepsilon = \varepsilon(\Omega) > 0$ with the following significance. Assume that $2 - \varepsilon < p < 2 + \varepsilon$, $0 \leq l \leq m$ and that the l-differential form $u \in L^p(\Omega, \Lambda^l T\mathcal{M})$ has, in the sense of distributions, $du \in L^p(\Omega, \Lambda^{l+1} T\mathcal{M})$ and $\delta u \in L^p(\Omega, \Lambda^{l-1} T\mathcal{M})$.*

Then the following are equivalent:

1. *$\nu \wedge u$, initially considered as in (11.10), actually belongs to $L^p(\partial\Omega, \Lambda^{l+1} T\mathcal{M})$;*
2. *$\nu \vee u$, initially considered as in (11.9), actually belongs to $L^p(\partial\Omega, \Lambda^{l-1} T\mathcal{M})$.*

Also, there exists a positive constant $C = C(\partial\Omega, p)$ such that whenever (1) and (2) above are fulfilled then

$$\begin{aligned} &\max\left\{\|\nu\wedge u\|_{L^p(\partial\Omega,\Lambda^{l+1}T\mathcal{M})}\,,\ \|\nu\vee u\|_{L^p(\partial\Omega,\Lambda^{l-1}T\mathcal{M})}\right\} \\ &\quad\le C\|u\|_{L^p(\Omega,\Lambda^l T\mathcal{M})}+C\|du\|_{L^p(\Omega,\Lambda^{l+1}T\mathcal{M})}+C\|\delta u\|_{L^p(\Omega,\Lambda^{l-1}T\mathcal{M})} \\ &\qquad + C\min\left\{\|\nu\wedge u\|_{L^p(\partial\Omega,\Lambda^{l+1}T\mathcal{M})}\,,\ \|\nu\vee u\|_{L^p(\partial\Omega,\Lambda^{l-1}T\mathcal{M})}\right\}, \end{aligned} \tag{11.13}$$

Furthermore, setting $s := 1/p$ if $p \ge 2$ and $s := 1 - 1/p$ if $p \le 2$, each of these conditions implies that

$$u \in \bigcap_{\mu>0} H^{s-\mu,p}(\Omega, \Lambda^l T\mathcal{M}) \tag{11.14}$$

and, for each $\mu > 0$, there exists $C = C(\mu, p, \Omega) > 0$ so that

$$\begin{aligned} &\|u\|_{H^{s-\mu,p}(\Omega,\Lambda^l T\mathcal{M})} \\ &\quad\le C\|u\|_{L^p(\Omega,\Lambda^l T\mathcal{M})}+C\|du\|_{L^p(\Omega,\Lambda^{l+1}T\mathcal{M})}+C\|\delta u\|_{L^p(\Omega,\Lambda^{l-1}T\mathcal{M})} \\ &\qquad + C\min\left\{\|\nu\wedge u\|_{L^p(\partial\Omega,\Lambda^{l+1}T\mathcal{M})}\,,\ \|\nu\vee u\|_{L^p(\partial\Omega,\Lambda^{l-1}T\mathcal{M})}\right\}. \end{aligned} \tag{11.15}$$

Finally, when $p = 2$, one can take $\mu = 0$ in (11.14) and (11.15).

A few comments are in order here. First, in the case when $p = 2$ (and, hence, $\mu = 0$) $u_{\rm tan} = 0$ or $u_{\rm nor} = 0$, and $\partial\Omega \in C^\infty$, estimates of the type (11.15) go back to Gaffney and Friedrichs ([Gaf], [Fr2]); in such a case $H^{1/2,2}$ may be replaced by $H^{1,2}$. However, in the class of Lipschitz domains the exponent $1/2$ is sharp (even for smooth metrics) as simple counterexamples show.

The equivalence (1) $\Leftrightarrow$ (2) for arbitrary $l \in \{0, 1, ..., m\}$ is sharp in the interval $2 \le p < \infty$. Indeed, for each $\lambda \in (0, 1)$ it is possible to construct a harmonic function v_λ in a cone like domain $\Omega_\lambda \subset \mathbb{R}^3$ with vertex at the origin, satisfying $\nabla_{\rm tan} v_\lambda \in L^\infty(\partial\Omega)$ and so that $|\nabla v_\lambda(x)| \approx |x|^{\lambda-1}$ as $|x| \to 0$. Then the 1-form $u_\lambda := dv_\lambda$ is closed, coclosed, satisfies $\nu \wedge u_\lambda \in L^\infty(\partial\Omega, \Lambda^2 T\mathcal{M})$ and $|\nu \vee u_\lambda(x)| \approx |x|^{\lambda-1}$ as $|x| \to 0$. Thus, $\nu \vee u_\lambda \in L^p(\partial\Omega)$ if and only if $p \notin (2/(1-\lambda), 3/(1-\lambda))$. Now the union of all such intervals when $\lambda \in (0, 1)$ is precisely $(2, \infty)$. In particular, for each $p > 2$ there exists a Lipschitz domain for which (2) $\Rightarrow$ (1) fails. A similar construction shows that (1) $\Rightarrow$ (2) also fails in the range $(2, \infty)$. It would be interesting to know whether the same is true for the range $(1, 2)$. We conjecture that this is indeed the case.

Second, in the light of the recent progress in [MT2] (cf. the last remark in §2), (11.4) can be sharpened to

$$u \in B^{p,p^*}_{1/p}(\Omega, \Lambda^l T\mathcal{M}).$$

The estimate (11.15) improves accordingly.

Third, the estimate (11.15) together with the classical Rellich selection lemma, trivially imply that the inclusion of the subspace of $L^p(\Omega, \Lambda^l T\mathcal{M})$ consisting of l-forms u with $du \in L^p(\Omega, \Lambda^{l+1}T\mathcal{M})$, $\delta u \in L^p(\Omega, \Lambda^{l-1}T\mathcal{M})$ and such that either (1) or (2) above holds, when equipped with the norm given by the right side of (11.15), embeds *compactly* into $L^p(\Omega, \Lambda^l T\mathcal{M})$. Such a result has found applications in electromagnetism and hydrodynamics; cf. [Te2], [MP], [Pi] for some related material.

Third, if $u \in L^p(\Omega, \Lambda^l T\mathcal{M})$ is so that $du \in L^p(\Omega, \Lambda^{l+1} T\mathcal{M})$ and $\nu \wedge u \in L^p(\partial\Omega, \Lambda^{l+1} T\mathcal{M})$ for some $1 < p < \infty$, then, necessarily, $\nu \wedge u$ is *normal*, i.e. $\nu \wedge u \in L^p_{\rm nor}(\partial\Omega, \Lambda^{l+1} T\mathcal{M})$.

PROOF. Select some fixed, constant, positive potential V on $\mathcal{M}$ and choose $\varepsilon = \varepsilon(\Omega, V) > 0$ so that the results in the Corollary 10.4 hold. Now, if $l = 0$ or $l = m$ the conclusions are trivial so we restrict attention to the range $1 \le l \le m-1$. Clearly, if we prove the implication "(1) $\Longrightarrow$ (2)" then the opposite one follows from this and an application of the Hodge star-isomorphism.

Hence, assume that u satisfies the hypotheses of the theorem and that, in addition, (1) holds true. Let ϕ be an arbitrary form in $C^1(\mathcal{M}, \Lambda^{l-1} T\mathcal{M})$. Since, by (6.16), $\nu \wedge \phi$ belongs to $L^{q,d}_{\rm nor}(\partial\Omega, \Lambda^l T\mathcal{M})$ with $1/p + 1/q = 1$, Corollary 10.4 ensures the existence of $g \in L^{q,d}_{\rm nor}(\partial\Omega, \Lambda^l T\mathcal{M})$ such that $(\frac{1}{2}I + N_l)g = \nu \wedge \phi$. Thus, if we now set $v := \delta\mathcal{S}_l g$ in Ω, then $v \in C^1(\Omega, \Lambda^l T\mathcal{M})$, $(\Delta - V)v = 0$ and $\delta v = 0$ in Ω. Moreover, from (6.17) we obtain $dv = -\delta\mathcal{S}_{l+1}(d_\partial g) - V\mathcal{S}_l g$. so that $\mathcal{N}(v)$, $\mathcal{N}(dv) \in L^q(\partial\Omega)$. These, integration by parts and an easy limiting argument then yield the identity

$$\int\!\!\int_\Omega \langle u, d(v - \phi)\rangle \, d\mathrm{Vol} = \int\!\!\int_\Omega \langle \delta u, v - \phi\rangle \, d\mathrm{Vol}. \tag{11.16}$$

Using this, (11.8) and repeated integrations by parts, we may write

$$\begin{aligned}
\langle\!\langle \nu \vee u, \phi \rangle\!\rangle &= -\int\!\!\int_\Omega \langle \delta u, \phi\rangle \, d\mathrm{Vol} + \int\!\!\int_\Omega \langle u, d\phi\rangle \, d\mathrm{Vol} \\
&= \int\!\!\int_\Omega \langle \delta u, v - \phi\rangle \, d\mathrm{Vol} - \int\!\!\int_\Omega \langle \delta u, v\rangle \, d\mathrm{Vol} \\
&\quad - \int\!\!\int_\Omega \langle u, d(v - \phi)\rangle \, d\mathrm{Vol} + \int\!\!\int_\Omega \langle u, dv\rangle \, d\mathrm{Vol} \\
&= \int\!\!\int_\Omega \langle u, dv\rangle \, d\mathrm{Vol} - \int\!\!\int_\Omega \langle \delta u, v\rangle \, d\mathrm{Vol} \\
&= -\int\!\!\int_\Omega \langle du, \mathcal{S}_{l+1}(d_\partial g)\rangle \, d\mathrm{Vol} + \langle\!\langle \nu \wedge u, \mathcal{S}_{l+1}(d_\partial g)\rangle\!\rangle \\
&\quad - \int\!\!\int_\Omega \langle u, V\mathcal{S}_l g\rangle \, d\mathrm{Vol} - \int\!\!\int_\Omega \langle \delta u, \delta\mathcal{S}_l g\rangle \, d\mathrm{Vol}.
\end{aligned} \tag{11.17}$$

Reconverting $\mathcal{S}_{l+1}(d_\partial g)$ into $d\mathcal{S}_l g$ (by (6.17), the last two identities in (10.2) are still valid in this case), (11.17), Schwarz's inequality and Theorem 2.9 allow us to derive the estimate

$$\begin{aligned}
&|\langle\!\langle \nu \vee u, \phi \rangle\!\rangle| \\
&\le C\|\nu \wedge u\|_{L^p(\partial\Omega, \Lambda^{l+1} T\mathcal{M})} \|g\|_{L^q(\partial\Omega, \Lambda^l T\mathcal{M})} \\
&\quad + C\left(\|du\|_{L^p(\Omega, \Lambda^{l+1} T\mathcal{M})} + \|\delta u\|_{L^p(\Omega, \Lambda^{l-1} T\mathcal{M})}\right) \|g\|_{L^q(\partial\Omega, \Lambda^l T\mathcal{M})}.
\end{aligned} \tag{11.18}$$

Since, from Corollary 10.4, we also have $\|g\|_{L^q(\partial\Omega, \Lambda^l T\mathcal{M})} \le C\|\phi\|_{L^q(\partial\Omega, \Lambda^{l-1} T\mathcal{M})}$, we may finally conclude from (11.18) that

$$\begin{aligned}
&|\langle\!\langle \nu \vee u, \phi \rangle\!\rangle| \\
&\le C\left(\|\delta u\|_{L^p(\Omega, \Lambda^{l-1} T\mathcal{M})} + \|du\|_{L^p(\Omega, \Lambda^{l+1} T\mathcal{M})}\right) \|\phi\|_{L^q(\partial\Omega, \Lambda^{l-1} T\mathcal{M})} \\
&\quad + C\|\nu \wedge u\|_{L^p(\partial\Omega, \Lambda^{l+1} T\mathcal{M})} \|\phi\|_{L^q(\partial\Omega, \Lambda^{l-1} T\mathcal{M})},
\end{aligned} \tag{11.19}$$

for any $\phi \in C^1(\mathcal{M}, \Lambda^{l-1}T\mathcal{M})$. Now Riesz's representation theorem implies that $\nu \vee u$ belongs to $L^p(\partial\Omega, \Lambda^{l-1}T\mathcal{M})$ plus a natural accompanying estimate. Thus we have (1) $\Rightarrow$ (2). As already noted, we hence have (2) $\Rightarrow$ (1), and (11.13) follows.

To prove the remaining part of the theorem we shall use a Green type representation formula for u. To begin with, assume that $u \in C^1(\overline{\Omega}, \Lambda^l T\mathcal{M})$. Fix V a constant, positive potential and denote by $\Gamma_l(x,y)$ the Schwartz kernel of the operator $(\Delta - V)^{-1}$ acting on (appropriate spaces of) l-forms. Then, successive integrations by parts against this kernel give that

$$\begin{aligned}
u(x) = & -\iint_\Omega \langle \delta_y \Gamma_l(x,y), (\delta u)(y)\rangle \, d\mathrm{Vol}(y) - \iint_\Omega \langle d_y \Gamma_l(x,y), (du)(y)\rangle \, d\mathrm{Vol}(y) \\
& - \iint_\Omega \langle \Gamma_l(x,y), (Vu)(y)\rangle \, d\mathrm{Vol}(y) + \int_{\partial\Omega} \langle d_y \Gamma_l(x,y), \nu(y) \wedge u(y)\rangle \, d\sigma(y) \\
& - \int_{\partial\Omega} \langle \delta_y \Gamma_l(x,y), \nu(y) \vee u(y)\rangle \, d\sigma(y) \\
= & - d\Pi_{l-1}(\delta u)(x) - \delta\Pi_{l+1}(du)(x) - \Pi_l(Vu)(x) \\
& - \iint_\Omega \langle Q_{l-1}(x,y), (\delta u)(y)\rangle \, d\mathrm{Vol}(y) - \iint_\Omega \langle R_l(x,y), (du)(y)\rangle \, d\mathrm{Vol}(y) \\
& + \delta\mathcal{S}_{l+1}(\nu \wedge u)(x) + R_l(\nu \wedge u)(x) - d\mathcal{S}_{l-1}(\nu \vee u)(x) - Q_{l-1}(\nu \vee u)(x),
\end{aligned} \tag{11.20}$$

for a.e. $x \in \Omega$. That (11.20) remains valid if u is only assumed to satisfy the hypotheses of the theorem as well as both (1) and (2), follows from the equality of weak and strong exterior derivatives (cf. K. O. Friedrichs [Fr1] and L. Hörmander [Hö]). Based on this identity, Theorem 2.9, Proposition 2.13 and (11.13), both (11.14) and (11.15) follow. The proof of Theorem 11.2 is finished. ■

We are now in a position to present the

PROOF OF (11.5). Consider some $u \in \mathbb{H}_V^{l,p}(\Omega)$ for $p \in (2-\varepsilon, 2+\varepsilon)$, with $\varepsilon > 0$ as in the conclusion of Theorem 11.2. Then Theorem 11.2 gives $\nu \wedge u \in L^q(\partial\Omega, \Lambda^{l+1}T\mathcal{M})$ for any $2-\varepsilon < q < 2+\varepsilon$. With this in hand, the identity (11.20), which remains valid in this context, and Proposition 1.1 yield $\mathcal{N}u \in L^q(\partial\Omega)$ for any $2-\varepsilon < q < 2+\varepsilon$. Thus $u \in \mathcal{H}_V^{l,q}(\Omega)$ for any $2-\varepsilon < q < 2+\varepsilon$. The desired conclusion then follows easily from this. ■

Finally, (11.6) is a corollary of the Hodge type decomposition contained in the next proposition. In this connection, let us point out that a quasi-conformal Yang-Mills theory on 4-dimensional manifolds (with applications to index theory) has been developed in [DS]. See also [IM], [Te1], [Pi] and [MP] for results related to this decomposition.

PROPOSITION 11.3. *With the notation and assumptions of Theorem 11.1, there holds the orthogonal decomposition*

$$L^2(\Omega, \Lambda^l T\mathcal{M}) = Im\, d_{l,2} \oplus Im\,(d_{l,2})^* \oplus \mathbb{H}_V^{l,2}(\Omega), \tag{11.21}$$

for each $l = 0, 1, ..., m$.

PROOF. We shall be brief since the proof is fairly standard once all ingredients are in place. First, plain functional analysis gives the weak Hodge decomposition $L^2(\Omega, \Lambda^l T\mathcal{M}) = \overline{\mathrm{Im}\, d_{l,2}} \oplus \overline{\mathrm{Im}\,(d_{l,2})^*} \oplus \mathbb{H}^l_\vee(\Omega)$ (cf., e.g., K. Kodaira in [Ko], p. 605 or Theorem 24 in [DR2]) so that we only need to show that all spaces involved are closed in L^2.

To this end, note that by (11.7) $\mathrm{Im}\, d_{l,p}$ has finite codimension in $\mathrm{Ker}\, d_{l,p}$ for each $1 < p < \infty$. Invoking the lemma below, it follows that $\mathrm{Im}\, d_{l,p}$ is closed in $\mathrm{Ker}\, d_{l,p}$ and, further, in $L^p(\Omega, \Lambda^l T\mathcal{M})$. Now Banach's closed range theorem gives a similar conclusion for $\mathrm{Im}\,(d_{l,p})^*$ and, for $p = 2$, this finishes the proof of the proposition (modulo that of Lemma 11.4). ■

Here is the lemma which ends the proof:

LEMMA 11.4. *Let $T : Dom\,(T) \subseteq X \to Y$ be a closed operator (in the sense of unbounded operators) between two Banach spaces and suppose that $Im T$ has finite codimension in Y. Then $Im T$ is a closed subspace of Y.*

PROOF. At least when T is everywhere defined (and, hence, bounded) this seems to be known; cf. Proposition 4.4.8, p. 167 in [Va]. The general case is easily reduced to this situation by considering T as a bounded operator from the Banach space $\mathrm{Dom}\,(T)$, equipped with the graph norm, into Y. ■

In closing let us point out that, by the remark made in the last part of § 10, in the case of C^1 domains the first part of Theorem 11.2 (i.e. the equivalence (1) $\Leftrightarrow$ (2)) is valid for the full range $1 < p < \infty$. In particular, if $\partial\Omega \in C^1$, then $\mathcal{H}^l_\vee(\Omega)$ and $\mathcal{H}^l_\wedge(\Omega)$ are independent of $p \in (1, \infty)$.

CHAPTER 12

The Proofs of the Theorems 5.1–5.5

Here we initiate the proofs of Theorems 5.1-5.5. As a preliminary matter, we first isolate several technical points in a series of lemmas, which we state at the beginning of this chapter, postponing their proof for the next chapter. We retain the same set of basic hypotheses on $\mathcal{M}$ as in the previous chapter.

Let Ω be a Lipschitz domain in $\mathcal{M}$ and let $V \in C^1(\mathcal{M})$ be a positive potential whose support intersects every connected component of $\mathcal{M}\setminus\bar{\Omega}$. Recall, from §6, the various layer potential operators introduced in connection with $\Delta - V$. Further, let $\varepsilon = \varepsilon(\Omega, V) > 0$ be sufficiently small so that the results of §§10 – 11 are valid for $2-\varepsilon < p < 2+\varepsilon$.

LEMMA 12.1. *Consider Ω as above and, for each $l \in \{0,1,...,m\}$, let T_l be the operator from $L^p_{\tan}(\partial\Omega, \Lambda^{l-1}T\mathcal{M}) \oplus L^p_{\tan}(\partial\Omega, \Lambda^l T\mathcal{M})$ into itself defined by*

$$T_l := \begin{pmatrix} -\frac{1}{2}I + \tilde{M}_{l-1} & \nu \vee S_l \\ -\nu \vee dQ_{l-1} & -\frac{1}{2}I + M_l \end{pmatrix}.$$

Then, if $2-\varepsilon < p < 2+\varepsilon$, it follows that

$$Im\ T_l = L^p_{\tan}(\partial\Omega, \Lambda^{l-1}T\mathcal{M}) \oplus \left\{\mathcal{H}^l_{\vee}(\Omega)|_{\partial\Omega}\right\}^{\circ}, \tag{12.1}$$

where the annihilator is taken in $L^p_{\tan}(\partial\Omega, \Lambda^l T\mathcal{M})$.

LEMMA 12.2. *Let Ω be a Lipschitz domain in $\mathcal{M}$ and $0 \le l \le m$. Suppose that the form $u \in C^1(\Omega, \Lambda^l T\mathcal{M})$ satisfies*

$$\Delta u = 0 \text{ in } \Omega, \text{ and } \mathcal{N}(u),\ \mathcal{N}(du) \in L^p(\partial\Omega). \tag{12.2}$$

Then, if $2-\varepsilon < p < 2+\varepsilon$, we have

$$\nu \vee (du)|_{\partial\Omega} \in \{\omega|_{\partial\Omega};\ \omega \in \mathcal{H}^l_{\vee}(\Omega)\}^{\circ}, \tag{12.3}$$

where the annihilator is taken in $L^p_{\tan}(\partial\Omega, \Lambda^l T\mathcal{M})$.

LEMMA 12.3. *Assume that Ω is a Lipschitz domain in $\mathcal{M}$ and $0 \le l \le m$. Then any form $u \in C^1(\Omega, \Lambda^l T\mathcal{M})$ satisfying*

$$\begin{cases} \Delta u = 0 \text{ in } \Omega, \\ \mathcal{N}(u), \mathcal{N}(du) \in L^p(\partial\Omega), \\ \nu \vee u|_{\partial\Omega} = 0, \end{cases} \tag{12.4}$$

for some $2-\varepsilon < p < 2+\varepsilon$, has the property that

$$\delta u = u_1 + u_2 \text{ in } \Omega,$$

for some

$$u_1 \in H^{1,p}(\Omega, \Lambda^{l-1}T\mathcal{M}) \tag{12.5}$$

and

(12.6)
$$u_2 \in C^0(\Omega, \Lambda^{l-1}T\mathcal{M}),\ \mathcal{N}(u_2) \in L^p(\partial\Omega) \text{ and } \exists\, u_2|_{\partial\Omega} \text{ pointwise a.e. on } \partial\Omega.$$

LEMMA 12.4. *For Ω a Lipschitz domain in $\mathcal{M}$, $2-\varepsilon < p < 2+\varepsilon$ and $0 \le l \le m$, there holds*

$$L^p_{\text{nor}}(\partial\Omega, \Lambda^l T\mathcal{M}) \cap \{\mathcal{H}^{l,q}(\Omega)|_{\partial\Omega}\}^\circ \subseteq L^{p,0}_{\text{nor}}(\partial\Omega, \Lambda^l T\mathcal{M}) \tag{12.7}$$

for $1/p + 1/q = 1$.

Of course, we also have

$$L^p_{\tan}(\partial\Omega, \Lambda^l T\mathcal{M}) \cap \{\mathcal{H}^{l,q}(\Omega)|_{\partial\Omega}\}^\circ \subseteq L^{p,0}_{\tan}(\partial\Omega, \Lambda^l T\mathcal{M}), \quad 1/p + 1/q = 1.$$

This follows from (12.7) and Hodge duality.

We devote §13 to the proofs of these four lemmas. Granted these results, we will now prove Theorems 5.1–5.5.

PROOF OF THEOREM 5.1. We treat in turn the nine parts of Theorem 5.1. The first order of business is to deal with the statement (1) regarding existence. The necessity of (5.8) is contained in Lemma 12.2. In order to prove that this is also sufficient, let g satisfy the compatibility condition (5.8) and fix a potential V as in the statement of Lemma 12.1. We express the candidate for a solution of $(BVP1)_l$ in the form

$$u(x) := \mathcal{S}_l k(x) + \int_{\partial\Omega} \langle \delta_y \Gamma_l(x,y), h(y)\rangle\, d\sigma(y), \quad x \in \Omega, \tag{12.8}$$

for $h \in L^p_{\tan}(\partial\Omega, \Lambda^{l-1}T\mathcal{M})$ and $k \in L^p_{\tan}(\partial\Omega, \Lambda^l T\mathcal{M})$ to be chosen shortly. Clearly, $\Delta u = 0$ in Ω and $\mathcal{N}(u) \in L^p(\partial\Omega)$. Also, since by (6.5)

$$du = d\mathcal{S}_l k - dQ_{l-1} h \text{ in } \Omega, \tag{12.9}$$

it follows that $\mathcal{N}(du) \in L^p(\partial\Omega)$ also. Based on (12.8) and (12.9), the two boundary conditions in $(BVP1)_l$ become

$$\begin{pmatrix} f \\ g \end{pmatrix} = \begin{pmatrix} \nu \vee u|_{\partial\Omega} \\ \nu \vee du|_{\partial\Omega} \end{pmatrix} = \begin{pmatrix} -\frac{1}{2}I + \tilde{M}_{l-1} & \nu \vee S_l \\ -\nu \vee dQ_{l-1} & -\frac{1}{2}I + M_l \end{pmatrix} \begin{pmatrix} h \\ k \end{pmatrix} = T_l \begin{pmatrix} h \\ k \end{pmatrix}. \tag{12.10}$$

The existence of h, k verifying (12.10) is now immediate from the right-to-left inclusion in (12.1) and this completes the proof of the statement (1).

Turning attention to (2), let u be a solution of $(BVP1)_l$ for $f = 0$, $g = 0$ and assume first that $2 \le p < 2 + \varepsilon$. Then the energy identity (8.8) gives $du = 0$, $\delta u = 0$ in Ω. Of course, for the integrations by parts leading up to (8.8) to work, appropriate control of δu in Ω is required *a priori*; indeed, (8.8) has been derived under the assumption that $\mathcal{N}(\delta u) \in L^2(\partial\Omega)$. However, inspection of the proof reveals that the slightly weaker regularity result on δu, furnished under the present circumstances by Lemma 12.3, suffices to conclude, much as before, that $du = 0$ and $\delta u = 0$ in Ω. Hence, the space of null solutions is precisely $\mathcal{H}^l_\vee(\Omega)$ which, by Theorem 11.1, has dimension $b_l(\Omega)$.

The case $2 - \varepsilon < p < 2$ can be reduced to the one just considered via a "bootstrapping" argument. The point is that, if u is as above, by utilizing the integral representation formula (13.10) in Lemma 13.1, one can show that the integrability properties of $\mathcal{N}(u)$ and $\mathcal{N}(du)$ improve. An iteration argument can then be used to conclude; we leave the details to the reader.

In particular, at this point we may conclude that any solution of $(BVP1)_l$ has the form

$$u = \mathcal{S}_l k + d\mathcal{S}_{l-1}h + Q_{l-1}h + \omega, \tag{12.11}$$

for some $h \in L^p_{\tan}(\partial\Omega, \Lambda^{l-1}T\mathcal{M})$, $k \in L^p_{\tan}(\partial\Omega, \Lambda^l T\mathcal{M})$ and $\omega \in \mathcal{H}^l_\vee(\Omega)$. Now (5.9) follows from this and Theorem 2.9. That $u \in H^{1/2,2}(\Omega, \Lambda^l T\mathcal{M})$ for $p = 2$, is also a consequence of (12.11) and Theorem 2.9.

Going further, we may employ (6.15) and Corollary 10.5 in conjunction with (12.10) to obtain that $h \in L^{p,\delta}_{\tan}(\partial\Omega, \Lambda^{l-1}T\mathcal{M})$ if $f \in L^{p,\delta}_{\tan}(\partial\Omega, \Lambda^{l-1}T\mathcal{M})$. Using this back in (12.11) and invoking (6.5), (6.17), it is clear that, in this case, $\mathcal{N}(\delta u) \in L^p(\partial\Omega)$. That, conversely, the latter condition forces $f \in L^{p,\delta}_{\tan}(\partial\Omega, \Lambda^{l-1}T\mathcal{M})$ is seen from the identity (6.15). Moreover, the estimate (5.10) follows from the discussion above and (12.11). Also, (3) applied to du proves (4).

Next, assume that $g = 0$ so that, by (4), $\mathcal{N}(\delta du) \in L^p(\partial\Omega)$. In turn, if $2 \le p < 2 + \varepsilon$, this can be used to justify the following integration by parts

$$\iint_\Omega |\delta du|^2 \, d\mathrm{Vol} = -\int_{\partial\Omega} \langle \nu \vee (du)|_{\partial\Omega}, (\delta du)|_{\partial\Omega} \rangle \, d\sigma = 0. \tag{12.12}$$

Hence, $\delta du = 0$ in Ω. Now, because

$$\iint_\Omega |du|^2 \, d\mathrm{Vol} = \iint_\Omega \langle \delta du, u \rangle \, d\mathrm{Vol} + \int_{\partial\Omega} \langle \nu \vee du, u \rangle \, d\sigma \tag{12.13}$$

we arrive at $du = 0$ in Ω as desired. When $2 - \varepsilon < p < 2$, a similar boot-strapping argument as before can be used, first for $v := du$ and then for u, in order to show that all integrands in (12.12)-(12.13) are still absolutely convergent; the same conclusion then follows.

The fact that we can prescribe periods (in which case the solution is unique) follows from what we have proved so far together with Theorem 11.1. This completes the proof of the statement (5).

Consider (6) now. In one direction, if $\delta du = 0$ in Ω then, by (6.15), g necessarily belongs to $L^{p,0}_{\tan}(\partial\Omega, \Lambda^l T\mathcal{M})$. Conversely, assume that $g \in L^{p,0}_{\tan}(\partial\Omega, \Lambda^l T\mathcal{M})$ so that, from (4), $\mathcal{N}(\delta du) \in L^p(\partial\Omega)$. Thus, $\nu \vee (d\delta u)|_{\partial\Omega} = -\nu \vee (\delta du)|_{\partial\Omega} = 0$ on $\partial\Omega$, so that du solves $(BVP1)_{l+1}$ for the pair of boundary data $(g, 0)$. Consequently, (5) yields $0 = d\delta u = \delta du$ and the proof of (6) is completed. Also, (7) is a consequence of all of the above and (6.15).

Suppose further that $\delta u = 0$ in Ω. Then, by (3) and (6), $f \in L^{p,0}_{\tan}(\partial\Omega, \Lambda^{l-1}T\mathcal{M})$ and $g \in L^{p,0}_{\tan}(\partial\Omega, \Lambda^l T\mathcal{M})$. Also, integrations by parts show that the membership of f to the space $\{\mathcal{H}^{l-1}_\vee(\Omega)|_{\partial\Omega}\}^\circ$ is a necessary condition.

Conversely, assuming that f, g are as in (5.12), we aim at proving that $\delta u = 0$ in Ω. First, from (6) and (3) we obtain $d\delta u = -\delta du = 0$ in Ω and $\mathcal{N}(\delta u) \in L^p(\partial\Omega)$ so that $\delta u \in \mathcal{H}^{l-1,p}_\vee(\Omega) = \mathcal{H}^{l-1,q}_\vee(\Omega)$, $1/p + 1/q = 1$, on account of (6.15) and Theorem 11.1. Using this, the fact that du is co-closed and the annihilation condition satisfied by f we have $\iint_\Omega |\delta u|^2 \, d\mathrm{Vol} = \int_{\partial\Omega} \langle f, \delta u \rangle \, d\sigma = 0$. Hence, $\delta u = 0$ in Ω as desired. The part in (8) concerning the problem $(BVP5)_l$ is obtained as before with the aid of Theorem 11.1.

Finally, (9) follows based on what we have already proved for $(BVP1)_l$ and the properties of the Hodge-$*$ isomorphism. The proof of Theorem 5.1 is finished. ■

Next, we turn our attention to the

PROOF OF THEOREM 5.2. First we show the necessity of the compatibility condition (5.20), i.e., that

$$\begin{aligned} &u \in C^1(\Omega, \Lambda^l T\mathcal{M}), \quad \Delta u = 0 \text{ in } \Omega, \quad \mathcal{N}(u), \mathcal{N}(du), \mathcal{N}(\delta u) \in L^p(\partial\Omega) \\ &\Longrightarrow \nu \vee (du)|_{\partial\Omega} - \nu \wedge (\delta u)|_{\partial\Omega} \in \{\mathcal{H}^{l,q}(\Omega)|_{\partial\Omega}\}^\circ, \end{aligned} \tag{12.14}$$

if $1/p + 1/q = 1$. To see this, let $(\Omega_j)_j$ be a sequence of nested subdomains of Ω, exhausting Ω in the usual way. Then, for each $\omega \in \mathcal{H}^{l,q}(\Omega)$, successive integrations by parts give

$$\begin{aligned} \int_{\partial\Omega} \langle \nu \vee du, \omega\rangle \, d\sigma &= \lim_{j\to\infty} \int_{\partial\Omega_j} \langle \nu_j \vee du, \omega\rangle \, d\sigma_j = -\lim_{j\to\infty} \int\!\!\int_{\Omega_j} \langle \delta du, \omega\rangle \, d\mathrm{Vol} \\ &= \lim_{j\to\infty} \int\!\!\int_{\Omega_j} \langle d\delta u, \omega\rangle \, d\mathrm{Vol} = \lim_{j\to\infty} \int_{\partial\Omega_j} \langle \nu_j \wedge \delta u, \omega\rangle \, d\sigma_j \\ &= \int_{\partial\Omega} \langle \nu \wedge \delta u, \omega\rangle \, d\sigma. \end{aligned}$$

This proves (12.14).

Conversely, assuming that $f - g$ satisfies the compatibility condition (5.20) we shall show that $(BVP7)_l$ has a solution. The key observation is that $u := v + w$ solves $(BVP7)_l$ provided

$$\begin{cases} v \in C^1(\Omega, \Lambda^l T\mathcal{M}), \\ \Delta v = 0 \text{ in } \Omega, \\ \mathcal{N}(v), \mathcal{N}(dv), \mathcal{N}(\delta v) \in L^p(\partial\Omega), \\ \nu \vee v|_{\partial\Omega} = 0, \\ \nu \vee (dv)|_{\partial\Omega} = f, \end{cases} \tag{12.15}$$

and

$$\begin{cases} w \in C^1(\Omega, \Lambda^l T\mathcal{M}), \\ \Delta w = 0 \text{ in } \Omega, \\ dw = 0 \text{ in } \Omega, \\ \mathcal{N}(w), \mathcal{N}(\delta w) \in L^p(\partial\Omega), \\ \nu \wedge w|_{\partial\Omega} = 0, \\ \nu \wedge (\delta w)|_{\partial\Omega} = g - \nu \wedge (\delta v)|_{\partial\Omega}. \end{cases} \tag{12.16}$$

Clearly, (12.15) is a particular case of $(BVP1)_l$ and it is not difficult to check that the boundary data satisfy the corresponding compatibility condition by virtue of (5.20) and the normality of g. Therefore, such a v exists. Going further, the problem (12.16) above corresponds to the dual under the Hodge star-isomorphism of $(BVP4)_l$ with boundary data 0 and $*(g - \nu \wedge \delta v)$. Consequently, it is solvable if (and only if) $g - \nu \wedge \delta v|_{\partial\Omega} \in L^{p,0}_{\mathrm{nor}}(\partial\Omega, \Lambda^l T\mathcal{M})$. However, since

$$g - \nu \wedge (\delta v)|_{\partial\Omega} = \nu \vee (dv)|_{\partial\Omega} - \nu \wedge (\delta v)|_{\partial\Omega} - (f - g) \in \{\mathcal{H}^{l,q}(\Omega)|_{\partial\Omega}\}^\circ, \tag{12.17}$$

by (12.14) and the compatibility condition (5.20), the desired conclusion is furnished by Lemma 12.4. This finishes the proof of the point (1) in Theorem 5.2.

Finally, if u is a solution of the homogeneous version of $(BVP7)_l$, then (8.8) readily implies that u is simultaneously closed and co-closed in Ω. The remaining parts in the statement of Theorem 5.2 are more or less direct consequences of the regularity results contained in Theorem 5.1. We omit the details. ■

We now present the

PROOF OF THEOREM 5.3. The necessity of (5.22) for the solvability of the problem $(BVP11)_l$ is clear from (6.15), (6.16). Conversely, let $f = \delta_\partial f'$, $g = d_\partial g'$, with $f' \in L^{p,\delta}_{\rm tan}(\partial\Omega, \Lambda^{l-1}T\mathcal{M})$, $g' \in L^{p,d}_{\rm nor}(\partial\Omega, \Lambda^{l+1}T\mathcal{M})$. Then $u := v + w$ is a solution to $(BVP11)_l$ provided v and w solve

$$
(12.18)\qquad \begin{cases} v \in C^1(\Omega, \Lambda^l T\mathcal{M}), \\ \Delta v = 0 \text{ in } \Omega, \\ dv = 0 \text{ in } \Omega, \\ \mathcal{N}(v),\, \mathcal{N}(\delta v) \in L^p(\partial\Omega), \\ \nu \vee v|_{\partial\Omega} = f', \end{cases}
$$

and

$$
(12.19)\qquad \begin{cases} w \in C^1(\Omega, \Lambda^l T\mathcal{M}), \\ \Delta w = 0 \text{ in } \Omega, \\ \delta w = 0 \text{ in } \Omega, \\ \mathcal{N}(w),\, \mathcal{N}(dw) \in L^p(\partial\Omega), \\ \nu \wedge w|_{\partial\Omega} = g', \end{cases}
$$

respectively. That (12.18) and (12.19) are solvable is guaranteed by Theorem 5.1; cf. especially $(BVP2)_l$ and its Hodge dual.

The last part in the statement of the theorem follows immediately from the fact that any harmonic field is a null solutions for $(BVP11)_l$. ■

Next, we present the

PROOF OF THEOREM 5.4. The necessity of (5.23) is once again clear from (6.16). For the opposite implication, let $g = d_\partial g'$ for some $g' \in L^{p,d}_{\rm nor}(\partial\Omega, \Lambda^{l+1}T\mathcal{M})$. We seek a solution of the form $u := v + w$ where v and w solve

$$
(12.20)\qquad \begin{cases} v \in C^1(\Omega, \Lambda^l T\mathcal{M}), \\ \Delta v = 0 \text{ in } \Omega, \\ \delta v = 0 \text{ in } \Omega, \\ \mathcal{N}(v),\, \mathcal{N}(dv) \in L^p(\partial\Omega), \\ \nu \wedge v|_{\partial\Omega} = g' \in L^{p,d}_{\rm nor}(\partial\Omega, \Lambda^{l+1}T\mathcal{M}), \end{cases}
$$

and

$$(12.21)\qquad \begin{cases} w \in C^1(\Omega, \Lambda^l T\mathcal{M}), \\ \Delta w = 0 \text{ in } \Omega, \\ dw = 0 \text{ in } \Omega, \\ \mathcal{N}(w) \in L^p(\partial\Omega), \\ \nu \vee w|_{\partial\Omega} = f - \nu \vee v|_{\partial\Omega} \in L^p_{\tan}(\partial\Omega, \Lambda^{l-1} T\mathcal{M}), \end{cases}$$

respectively. As before, the solvability of (12.20) and (12.21) is part of Theorem 5.1.

To see that the space of null solutions for $(BVP12)_l$ is finite dimensional, consider the mapping assigning du to each null solution u. Its image is contained in $\mathcal{H}^{l+1}_\wedge(\Omega)$ and it is not difficult to check that its kernel is precisely $\mathcal{H}^l_\vee(\Omega)$. Hence, by the results in §11 and linear algebra, the dimension of the space of null solutions for $(BVP12)_l$ is $\leq b_l(\Omega) + b_{m-l-1}(\Omega)$. ■

We conclude this chapter with the

PROOF OF THEOREM 5.5. The necessity of (5.24) follows from $(BVP14)_l$, a limiting argument and repeated integrations by parts; we omit the details. As for sufficiency, assume that (5.24) holds for each ω satisfying (5.25). Our strategy is to look for a solution u of $(BVP14)_l$ in the form $u := v + w$ where v and w solve

$$(12.22)\qquad \begin{cases} v \in C^1(\Omega, \Lambda^l T\mathcal{M}), \\ \Delta v = 0 \text{ in } \Omega, \\ \mathcal{N}(v), \mathcal{N}(dv), \mathcal{N}(\delta v) \in L^p(\partial\Omega), \\ \nu \vee v|_{\partial\Omega} = 0 \text{ on } \Omega, \\ \nu \vee (dv)|_{\partial\Omega} = g \in L^p_{\tan}(\partial\Omega, \Lambda^l T\mathcal{M}), \end{cases}$$

and

$$(12.23)\qquad \begin{cases} w \in C^1(\Omega, \Lambda^l T\mathcal{M}), \\ dw = 0 \text{ in } \Omega, \\ \delta w = 0 \text{ in } \Omega, \\ \mathcal{N}(w) \in L^p(\partial\Omega), \\ \nu \wedge w|_{\partial\Omega} = f - \nu \wedge v|_{\partial\Omega} \in L^p_{\mathrm{nor}}(\partial\Omega, \Lambda^{l+1} T\mathcal{M}), \end{cases}$$

respectively. That (12.22) is indeed solvable follows from (5.24) and Theorem 5.1. There remains to check that (12.23) is also solvable. By Theorem 5.1, this amounts to checking that

$$(12.24)\qquad f - \nu \wedge v|_{\partial\Omega} \in L^{p,0}_{\mathrm{nor}}(\partial\Omega, \Lambda^{l+1} T\mathcal{M}) \cap \left[\mathcal{H}^{l+1}_\wedge(\Omega)\right]^\circ$$

so that, thanks to Lemma 12.4, it suffices to show that

$$(12.25)\qquad f - \nu \wedge v|_{\partial\Omega} \in \left[\mathcal{H}^{l+1,q}(\Omega)\right]^\circ, \qquad 1/p + 1/q = 1.$$

To this end, consider an arbitrary $h \in \mathcal{H}^{l+1,q}(\Omega)$ and let ω solve

$$
\left\{
\begin{array}{l}
\omega \in C^1(\Omega, \Lambda^l T\mathcal{M}), \\
\Delta\omega = 0 \text{ in } \Omega, \\
\delta\omega = 0 \text{ in } \Omega, \\
\mathcal{N}(\omega), \mathcal{N}(d\omega) \in L^q(\partial\Omega), \\
\nu \vee \omega|_{\partial\Omega} = 0 \text{ on } \partial\Omega, \\
\nu \vee (d\omega)|_{\partial\Omega} = \nu \vee h|_{\partial\Omega} \in L^{q,0}_{\tan}(\partial\Omega, \Lambda^l T\mathcal{M}).
\end{array}
\right.
\tag{12.26}
$$

Again, the fact that this problem is solvable is a consequence of Theorem 5.1. Thus,

$$
\begin{aligned}
\int_{\partial\Omega} \langle f - \nu \wedge v, h \rangle \, d\sigma &= \int_{\partial\Omega} \langle f - \nu \wedge v, d\omega \rangle \, d\sigma \\
&= \int_{\partial\Omega} \langle f, d\omega \rangle \, d\sigma - \int_{\partial\Omega} \langle g, \omega \rangle \, d\sigma = 0.
\end{aligned}
\tag{12.27}
$$

The first equality is seen from the fact that $f - \nu \wedge v$ is normal and the last boundary condition in (12.26). Going further, the second equality is a consequence of (12.22) and integrations by parts, while the last equality follows directly from the compatibility condition (5.24). This proves (12.25) and, hence, (12.24). In turn, this implies that (12.23) is solvable and, further, that $(BVP14)_l$ is solvable too.

The remaining claims in the statement of Theorem 5.5 are straightforward consequences of what we have proved so far and we leave the details to the interested reader. ■

CHAPTER 13

The Proofs of the Auxiliary Lemmas

In this chapter we conclude the proofs of Theorems 5.1–5.5 initiated in §12 by presenting the proofs of Lemmas 12.1–12.4 stated there. We debut with the

PROOF OF LEMMA 12.1. Recall from (12.10) that for $h \in L^p_{\rm tan}(\partial\Omega, \Lambda^{l-1}T\mathcal{M})$ and $k \in L^p_{\rm tan}(\partial\Omega, \Lambda^l T\mathcal{M})$ we have $T_l(h,k) = (\nu \vee u, \nu \vee du)$, where u is defined as in (12.8). Consequently, the left-to-right inclusion in (12.1) is contained in Lemma 12.2.

In order to prove the opposite inclusion observe first that from Theorem 6.2 and Corollary 10.3 the operator T_l is Fredholm (with index zero). Thus, since T_l has closed range, it suffices to prove

$$\text{Ker}\, T_l^* \subseteq 0 \oplus \{\omega|_{\partial\Omega};\ \omega \in \mathcal{H}^l_V(\Omega)\}; \tag{13.1}$$

here T_l^*, mapping $L^q_{\rm tan}(\partial\Omega, \Lambda^{l-1}T\mathcal{M}) \oplus L^q_{\rm tan}(\partial\Omega, \Lambda^l T\mathcal{M})$ into itself, $1/p + 1/q = 1$, is the adjoint of T_l. Based on (6.13) and (6.14) it follows that the membership of (h,k) to $\text{Ker}\, T_l^*$ is equivalent with the system of boundary integral equations

$$\begin{cases} (-\frac{1}{2}I + N_l)(\nu \wedge h) = -\nu \wedge \displaystyle\int_{\partial\Omega} \langle d_y R_{l-1}(\cdot, y), \nu(y) \wedge k(y)\rangle\, d\sigma(y), \\ (-\frac{1}{2}I + \tilde{N}_{l+1})(\nu \wedge k) = -\nu \wedge \displaystyle\int_{\partial\Omega} \langle \Gamma_l(\cdot, y), \nu(y) \wedge h(y)\rangle\, d\sigma(y). \end{cases} \tag{13.2}$$

On account of (6.16) and Corollary 10.5 we see from (13.2) that in fact both $\nu \wedge h$ and $\nu \wedge k$ are "regular", i.e. $\nu \wedge h \in L^{2,d}_{\rm nor}(\partial\Omega, \Lambda^l T\mathcal{M})$ and $\nu \wedge k \in L^{2,d}_{\rm nor}(\partial\Omega, \Lambda^{l+1}T\mathcal{M})$.

Going further, define

$$v := \mathcal{S}_l(\nu \wedge h) + \int_{\partial\Omega} \langle d_y \Gamma_l(\cdot, y), \nu(y) \wedge k(y)\rangle\, d\sigma(y) \quad \text{in } \mathcal{M} \setminus \partial\Omega. \tag{13.3}$$

Then $(\Delta - V)v = 0$ in $\mathcal{M} \setminus \partial\Omega$. Also, using the regularity of $\nu \wedge h$ and $\nu \wedge k$ and arguing as before it follows that $\mathcal{N}(v), \mathcal{N}(dv), \mathcal{N}(\delta v) \in L^2(\partial\Omega)$. Furthermore, since $\delta_x d_y \Gamma_l(x,y) = -\delta_x R_l(x,y) = d_y R_{l-1}(x,y)$, we get that

$$\nu \wedge (\delta v)|_{\partial\Omega_-} = (-\tfrac{1}{2}I + N_l)(\nu \wedge h) + \nu \wedge \int_{\partial\Omega} \langle d_y R_{l-1}(\cdot, y), \nu(y) \wedge k(y)\rangle\, d\sigma(y) = 0. \tag{13.4}$$

Also, from the second equality in (13.2), we see that

$$\nu \wedge v|_{\partial\Omega_-} = (-\tfrac{1}{2}I + \tilde{N}_{l+1})(\nu \wedge k) + \nu \wedge \int_{\partial\Omega} \langle \Gamma_l(\cdot, y), \nu(y) \wedge h(y)\rangle\, d\sigma(y) = 0. \tag{13.5}$$

Inserting (13.4)–(13.5) in the energy identity (8.8) gives that $dv = 0$, $\delta v = 0$ in Ω_-, $v = 0$ on $\text{supp}\, V$. Hence, invoking Theorem 4.3, we arrive at $v = 0$ in Ω_-. In turn, this and the jump-relations derived in §6 imply $\nu \vee v|_{\partial\Omega_+} = \nu \vee v|_{\partial\Omega_-} = 0$ and

$\nu \vee (dv)|_{\partial\Omega_+} = \nu \vee (dv)|_{\partial\Omega_-} = 0$. These identities and the energy identity (8.8) on Ω force both δv and dv to vanish in Ω so that, in particular, $v|_\Omega \in \mathcal{H}^l_\vee(\Omega)$. Now,

$$\nu \wedge h = \nu \wedge (\delta v)|_{\partial\Omega_+} - \nu \wedge (\delta v)|_{\partial\Omega_-} = 0, \tag{13.6}$$

whereas

$$k = \nu \vee (\nu \wedge v|_{\partial\Omega_+} - \nu \wedge v|_{\partial\Omega_-}) = v|_{\partial\Omega} \in \mathcal{H}^l_\vee(\Omega)|_{\partial\Omega}. \tag{13.7}$$

With this, the proof of Lemma 12.1 is completed (modulo the proof of Lemma 12.2). ■

Let us note that, as a consequence of (11.3) and the fact that T_l has index zero, we have

$$\dim \operatorname{Ker} T_l = b_l(\Omega), \quad \text{the } l\text{-th Betti number of } \Omega.$$

This is a remarkable identity in as much as it relates a purely analytical object (the operator T_l) to a purely topological entity (a Betti number). In the presence of irregularities such connections are considerably less understood.

The fundamental role played by the topology is further exhibited in the following result. First, since $\operatorname{index} T_l = 0$, we have that

$$T_l \text{ is invertible} \iff b_l(\Omega) = 0.$$

In concert with (12.10), this implies that, for $2-\varepsilon < p < 2+\varepsilon$, the two statements below are equivalent:

1. there exists $C > 0$ such that

$$\|u\|_{L^p(\partial\Omega,\Lambda^l T\mathcal{M})} \leq C\left(\|\nu \vee u\|_{L^p(\partial\Omega,\Lambda^{l-1}T\mathcal{M})} + \|\nu \vee du\|_{L^p(\partial\Omega,\Lambda^l T\mathcal{M})}\right)$$

   for any l-form u harmonic in Ω and having $\mathcal{N}(u)$, $\mathcal{N}(du) \in L^p(\partial\Omega)$;
2. the l-th Betti number $b_l(\Omega)$ vanishes.

This can be regarded as a sharp form of the main result in [IMS].

Next, we turn to the

PROOF OF LEMMA 12.2. Let $u \in C^1(\Omega, \Lambda^l T\mathcal{M})$ be as in (12.2), and let v be an arbitrary solution of

$$\begin{cases} v \in C^1(\Omega, \Lambda^l T\mathcal{M}), \\ \Delta v = 0 \text{ in } \Omega, \\ \mathcal{N}(v), \mathcal{N}(dv) \in L^p(\partial\Omega), \\ \nu \vee v|_{\partial\Omega} = \nu \vee u|_{\partial\Omega}, \\ \nu \vee (dv)|_{\partial\Omega} = 0. \end{cases} \tag{13.8}$$

The important thing is that the existence of such a v is guaranteed by the right-to-left inclusion in (12.1) (this particular inclusion has an independent proof which has been presented above).

Now, the difference $w := u - v$ enjoys the same properties as u and, in addition, $\nu \vee w|_{\partial\Omega} = 0$. Thus, δw can be written as in Lemma 12.3. In turn, this suffices to justify, via a limiting argument involving an approximating sequence $\Omega_j \nearrow \Omega$, the following integration by parts, for any $\omega \in \mathcal{H}^l_\vee(\Omega)$,

$$\int_{\partial\Omega} \langle \nu \vee dw, \omega \rangle \, d\sigma = - \iint_\Omega \langle \delta dw, \omega \rangle \, d\mathrm{Vol} = \iint_\Omega \langle d\delta w, \omega \rangle \, d\mathrm{Vol}$$
$$= \int_{\partial\Omega} \langle \delta w, \nu \vee \omega \rangle \, d\sigma = 0.$$

Hence, $\nu \vee (dw)|_{\partial\Omega} \in \{\mathcal{H}^l_\vee(\Omega)|_{\partial\Omega}\}^\circ$. Now, since by (13.8) we have $\nu \vee (du)|_{\partial\Omega} = \nu \vee (dw)|_{\partial\Omega}$, the proof of Lemma 12.2 is completed (modulo the proof of Lemma 12.3). ∎

Lemma 12.3 is a consequence of the remarkable integral representation formula contained in the lemma below. To state it, recall the trace operator Tr in the Sobolev space and the volume potential operator Π_l from (11.11).

LEMMA 13.1. *Let Ω be a Lipschitz domain in $\mathcal{M}$ and fix a constant, strictly positive potential V on $\mathcal{M}$. Also, let $\varepsilon = \varepsilon(\Omega, V) > 0$ be chosen so that the results in §§ 10 – 11 hold for $2 - \varepsilon < p < 2 + \varepsilon$.*

Assume that, for some $0 \leq l \leq m$, the harmonic form $u \in C^1(\Omega, \Lambda^l T\mathcal{M})$ satisfies

$$\mathcal{N}(u), \mathcal{N}(du) \in L^p(\partial\Omega) \text{ and } \nu \vee u|_{\partial\Omega} = 0 \tag{13.9}$$

for some $2 - \varepsilon < p < 2 + \varepsilon$. Then u has the following integral representation in Ω:
(13.10)

$$\begin{aligned} u = &- V\Pi_l u + V\delta\mathcal{S}_{l+1}\left[\left(-\tfrac{1}{2}I + N_{l+1}\right)^{-1}(\nu \wedge Tr(\Pi_l u))\right] \\ &+ V\mathcal{S}_l\left[\left(-\tfrac{1}{2}I + N_l\right)^{-1}(\nu \wedge Tr(\delta\Pi_l u))\right] \\ &- V\delta\mathcal{S}_{l+1}\left[\left(-\tfrac{1}{2}I + N_{l+1}\right)^{-1}\left[\nu \wedge \mathcal{S}_l\left(\left(-\tfrac{1}{2}I + N_l\right)^{-1}(\nu \wedge Tr(\delta\Pi_l u))\right)\right]\right] \\ &- \mathcal{S}_l\left(\nu \vee (du)|_{\partial\Omega}\right) + \delta\mathcal{S}_{l+1}\left(\left(-\tfrac{1}{2}I + N_{l+1}\right)^{-1}(\nu \wedge \mathcal{S}_l(\nu \vee (du)|_{\partial\Omega}))\right) \\ &+ \mathcal{S}_l\left(\left(-\tfrac{1}{2}I + N_l\right)^{-1}(\nu \wedge \delta\mathcal{S}_l(\nu \vee (du)|_{\partial\Omega}))\right) \\ &- \delta\mathcal{S}_{l+1}\left[\left(-\tfrac{1}{2}I + N_{l+1}\right)^{-1}\left[\nu \wedge \mathcal{S}_l\left(\left(-\tfrac{1}{2}I + N_l\right)^{-1}\nu \wedge \delta\mathcal{S}_l(\nu \vee (du)|_{\partial\Omega})\right)\right]\right]. \end{aligned}$$

PROOF. Let $\Omega_j \nearrow \Omega$ be a sequence of Lipschitz domains approximating Ω as in the last part of the proof of Theorem 3.1. Now, for each j, set

$$U_{l,j}(x,y) := \left(\mathrm{Id} \otimes (-\tfrac{1}{2}I + M_{l,j})^{-1}(\nu_j \vee \mathrm{Tr}_j \, d)\right)\Gamma_l(x,y) \tag{13.11}$$

for $x \in \Omega_j$, $y \in \partial\Omega_j$, and introduce the Green type function $G_{l,j}$ by appropriately adjusting the fundamental solution Γ_l, i.e.,

$$\begin{aligned} G_{l,j}(x,y) := \Gamma_l(x,y) &- (\mathrm{Id} \otimes \mathcal{S}_{l,j})\, U_{l,j}(x,y) \\ &- \left[\mathrm{Id} \otimes \left(d\mathcal{S}_{l-1,j}((-\tfrac{1}{2}I + M_{l-1,j})^{-1}\nu_j \vee \mathrm{Tr}_j)\right)\right]\Gamma_l(x,y) \\ &+ \left[\mathrm{Id} \otimes \left(d\mathcal{S}_{l-1,j}((-\tfrac{1}{2}I + M_{l-1,j})^{-1}\nu_j \vee \mathcal{S}_{l,j})\right)\right] U_{l,j}(x,y) \end{aligned} \tag{13.12}$$

for $x, y \in \Omega_j \times \Omega_j \setminus \mathrm{diag}$. Hereafter, $M_{l,j}$, $\mathcal{S}_{l,j}$, etc., are the analogue of M_l, $\mathcal{S}_l$, etc., constructed in connection with Ω_j.

Note that the correction terms are annihilated by $\mathrm{Id}\otimes(\Delta-V)$ and, hence, $G_{l,j}(x,\cdot)$ retains the character of being a fundamental solution for $\mathrm{Id}\otimes(\Delta-V)$ with mass at x. Other properties that will play an important role in subsequent calculations are summarized below:

$$
(13.13)\qquad \begin{cases} \nu_j(y)\vee G_{l,j}(x,y)|_{\partial\Omega_j}=0 \text{ at a.e. } y\in\partial\Omega_j,\\ \nu_j(y)\vee d_yG_{l,j}(x,y)|_{\partial\Omega_j}=0 \text{ at a.e. } y\in\partial\Omega_j,\\ \mathcal{N}_j(G_{l,j}(x,\cdot)),\,\mathcal{N}_j(dG_{l,j}(x,\cdot)),\,\mathcal{N}_j(\delta G_{l,j}(x,\cdot))\in L^q(\partial\Omega_j),\\ \sup\limits_j\|\mathcal{N}_j(G_{l,j}(x,\cdot))\|_{L^q(\partial\Omega_j)},\ \sup\limits_j\|\mathcal{N}_j(\delta G_{l,j}(x,\cdot))\|_{L^q(\partial\Omega_j)}<+\infty,\end{cases}
$$

for $1/p+1/q=1$. The first two equalities are easily checked from the explicit form of $G_{l,j}$, while the third line follows from Corollary 10.4 and (6.17). Finally, the last line of (13.13) is a consequence of the uniform invertibility of the operators $-\frac{1}{2}I+M_{l,j}$ on $L^{q,\delta}_{\tan}(\partial\Omega_j,\Lambda^lT\mathcal{M})$ with respect to j, i.e.,

$$
(13.14)\qquad \sup_j\|(-\tfrac{1}{2}I+M_{l,j})^{-1}\|<+\infty.
$$

In turn, for $q=2$, (13.14) can be proved from the fact that $(\Omega_j)_j$ have bounded Lipschitz character in j and the estimates in Theorem 8.2. Th extension to $2-\varepsilon<q<2+\varepsilon$ then follows from this and the results in [KM].

Going further, if u is an arbitrary smooth l-form, repeated applications of (4.8) give that

$$
(13.15)\qquad \begin{aligned} u(x)=&\iint_{\Omega_j}\langle(\Delta-V)u\,,\,G_{l,j}(x,\cdot)\rangle\,d\mathrm{Vol}\\ &-\int_{\partial\Omega_j}\langle\nu_j\vee du,G_{l,j}(x,\cdot)\rangle\,d\sigma_j+\int_{\partial\Omega_j}\langle\delta u,\nu_j\vee G_{l,j}(x,\cdot)\rangle\,d\sigma_j\\ &+\int_{\partial\Omega_j}\langle u,\nu_j\vee dG_{l,j}(x,\cdot)\rangle\,d\sigma_j-\int_{\partial\Omega_j}\langle\nu_j\vee u,\delta G_{l,j}(x,\cdot)\rangle\,d\sigma_j\end{aligned}
$$

for each fixed $x\in\Omega$ and each j (large enough). Observe now that by (13.13) the second and the third boundary integrals above vanish so that, if u is also harmonic, (13.15) reduces to

$$
(13.16)\qquad \begin{aligned} u(x)=&-V\iint_{\Omega_j}\langle u\,,\,G_{l,j}(x,\cdot)\rangle\,d\mathrm{Vol}-\int_{\partial\Omega_j}\langle\nu_j\vee du,G_{l,j}(x,\cdot)\rangle\,d\sigma_j\\ &-\int_{\partial\Omega_j}\langle\nu_j\vee u,\delta G_{l,j}(x,\cdot)\rangle\,d\sigma_j.\end{aligned}
$$

Let us first deal with the solid integral above. The idea is to use the explicit expression of $G_{l,j}$ and to transfer progressively (by passing to adjoints) all the operators acting on Γ_l to the other differential form participating in the pairing. In doing so, the following transposition formulas are useful

$$
(13.17)\qquad \begin{aligned} &\iint_\Omega\langle v,(\mathrm{Id}\otimes d\mathcal{S}_l)\omega\rangle\,d\mathrm{Vol}=\int_{\partial\Omega}\langle\omega,\mathrm{Tr}\,(\Pi_lv)\rangle\,d\sigma,\\ &\iint_\Omega\langle v,(\mathrm{Id}\otimes d\mathcal{S}_l)\omega\rangle\,d\mathrm{Vol}=\int_{\partial\Omega}\langle\omega,\mathrm{Tr}\,(\delta\Pi_lv)\rangle\,d\sigma.\end{aligned}
$$

In (13.17) both the differential form v as well as the double form ω are assumed to be, e.g., continuous on the closure of the Lipschitz domain Ω, and have appropriate degrees so that the pairings make sense. Based on these and (6.13), a calculation gives

(13.18)
$$\begin{aligned}
&\iint_{\Omega_j} \langle u\,,\, G_{l,j}(x,\cdot)\rangle \, d\mathrm{Vol}\\
&= (\Pi_{l,j}u)(x) - \delta\mathcal{S}_{l+1,j}\Big[\big(-\tfrac{1}{2}I + N_{l+1,j}\big)^{-1}(\nu_j \wedge \mathrm{Tr}_j\,(\Pi_{l,j}u))\Big](x)\\
&- \mathcal{S}_{l,j}\Big[\big(-\tfrac{1}{2}I + N_{l,j}\big)^{-1}(\nu_j \wedge \mathrm{Tr}_j\,(\delta\Pi_{l,j}u))\Big](x)\\
&+ \delta\mathcal{S}_{l+1,j}\Big[\big(-\tfrac{1}{2}I + N_{l+1,j}\big)^{-1}\Big[\nu_j \wedge \mathcal{S}_{l,j}\Big(\big(-\tfrac{1}{2}I + N_{l,j}\big)^{-1}(\nu_j \wedge \mathrm{Tr}_j\,(\delta\Pi_{l,j}u))\Big)\Big]\Big](x).
\end{aligned}$$

In a similar fashion,

(13.19)
$$\begin{aligned}
&\int_{\partial\Omega_j} \langle \nu_j \vee du, G_{l,j}(x,\cdot)\rangle \, d\sigma_j = -\mathcal{S}_{l,j}\Big(\big(-\tfrac{1}{2}I + N_{l,j}\big)^{-1}(\nu_j \wedge \delta\mathcal{S}_{l,j}(\nu_j \vee (du)|_{\partial\Omega_j}))\Big)\\
&+ \mathcal{S}_{l,j}\,(\nu_j \vee (du)|_{\partial\Omega_j}) - \delta\mathcal{S}_{l+1,j}\Big(\big(-\tfrac{1}{2}I + N_{l+1,j}\big)^{-1}(\nu_j \wedge \mathcal{S}_{l,j}(\nu_j \vee (du)|_{\partial\Omega_j}))\Big)\\
&+ \delta\mathcal{S}_{l+1,j}\Big\{\big(-\tfrac{1}{2}I + N_{l+1,j}\big)^{-1}\\
&\qquad \Big[\nu_j \wedge \mathcal{S}_{l,j}\Big(\big(-\tfrac{1}{2}I + N_{l,j}\big)^{-1}\nu_j \wedge \delta\mathcal{S}_{l,j}(\nu_j \vee (du)|_{\partial\Omega_j})\Big)\Big]\Big\}.
\end{aligned}$$

Finally, by (13.10), the last condition in (13.13) and Lebesgue's dominated convergence theorem we have

(13.20)
$$\left|\int_{\partial\Omega_j} \langle \nu_j \vee u, \delta G_{l,j}(x,\cdot)\rangle \, d\sigma_j\right| \leq C\|\nu_j \vee u\|_{L^p(\partial\Omega_j,\Lambda^{l-1}T\mathcal{M})} \longrightarrow 0 \text{ as } j \to \infty.$$

In summary, combining (13.18), (13.19) and passing to the limit, the identity (13.10) follows on account of (13.20). This completes the proof. ■

Proof of Lemma 12.3. If u is as in (12.4), then u can be expressed as in (13.10) for some constant, positive potential V. The crucial fact is that $\delta\delta = 0$ so that $\delta u = u_1 + u_2$ with

$$\begin{aligned}
u_1 &:= -V\delta\Pi_l u,\\
u_2 &:= V\delta\mathcal{S}_l\Big[\big(-\tfrac{1}{2}I + N_l\big)^{-1}(\nu \wedge \mathrm{Tr}\,(\delta\Pi_l u))\Big] - \delta\mathcal{S}_l\,(\nu \vee (du)|_{\partial\Omega})\\
&\quad + \delta\mathcal{S}_l\Big(\big(-\tfrac{1}{2}I + N_l\big)^{-1}(\nu \wedge \delta\mathcal{S}_l(\nu \vee (du)|_{\partial\Omega}))\Big).
\end{aligned}$$

By invoking Theorem 2.9 (cf. (11.12)), it clearly follows that u_1 and u_2 satisfy (12.5) and (12.6), respectively. ■

Finally, we present the

Proof of Lemma 12.4. Let g be an arbitrary form in $L^p_{\mathrm{nor}}(\partial\Omega, \Lambda^l T\mathcal{M}) \cap \{\mathcal{H}^{l,q}(\Omega)|_{\partial\Omega}\}^{\circ}$, $1/p + 1/q = 1$. We aim at proving that $d_{\partial}g = 0$. To this end, let

$\psi \in C^1(\bar{\Omega}, \Lambda^{l+1}T\mathcal{M})$ be arbitrary. The key idea is to invoke Theorem 5.1 (whose proof did not use Lemma 12.4) to guarantee the existence of a form u satisfying

$$\text{(13.21)} \qquad \begin{cases} u \in C^1(\Omega, \Lambda^l T\mathcal{M}), \\ du = 0 \text{ in } \Omega, \\ \delta u = 0 \text{ in } \Omega, \\ \mathcal{N}(u) \in L^q(\partial\Omega), \\ \nu \vee u|_{\partial\Omega} = \nu \vee (\delta\psi)|_{\partial\Omega}. \end{cases}$$

Notice that the compatibility condition $\nu \vee (\delta\psi)|_{\partial\Omega} \in \left\{\mathcal{H}_{\vee}^{l-1}(\Omega)\big|_{\partial\Omega}\right\}^0$ is easily checked integrating by parts. With this at hand, we may write

$$\int_{\partial\Omega} \langle g, \delta\psi\rangle \, d\sigma = \int_{\partial\Omega} \langle \nu \vee g, \nu \vee u\rangle \, d\sigma = \int_{\partial\Omega} \langle g, u|_{\partial\Omega}\rangle \, d\sigma = 0,$$

since $u \in \mathcal{H}^{l,q}(\Omega)$. This forces $d_\partial g = 0$ and, hence, finishes the proof of Lemma 12.4. ■

CHAPTER 14

Applications to Maxwell's Equations in Lipschitz Domains

In this chapter we shall indicate how the classical time-harmonic Maxwell system, considered here on arbitrary Lipschitz subdomains of Riemannian manifolds, can be subsumed within the theory we have developed so far. As such, this is a generalization of the results in [Mi2], [Mi3], [JM] which deal with bounded Lipschitz domains in the flat, Euclidean setting.

The departure point is to treat the analogue of $(BVP6)_l$ in § 5 written for the Helmholtz operator $\Delta + k^2$, $k \in \mathbb{C}$. In this regard, we shall prove the following.

THEOREM 14.1. *Let the Riemannian manifold $\mathcal{M}$ be as in § 5 and let $\Omega \subset \mathcal{M}$ be an arbitrary Lipschitz domain. For $k \in \mathbb{C}$ and $l \in \{0, 1, ..., m\}$, consider the boundary value problem*

$$(14.1)\qquad (BVP_k)_l \begin{cases} u \in C^1(\Omega, \Lambda^l T_{\mathbb{C}}\mathcal{M}), \\ (\Delta + k^2)u = 0 \text{ in } \Omega, \\ \mathcal{N}(u), \mathcal{N}(\delta u) \in L^p(\partial\Omega), \\ \nu \wedge u\big|_{\partial\Omega} = f \in L^p_{\rm nor}(\partial\Omega, \Lambda^{l+1}T_{\mathbb{C}}\mathcal{M}), \\ \nu \wedge (\delta u)\big|_{\partial\Omega} = g \in L^p_{\rm nor}(\partial\Omega, \Lambda^l T_{\mathbb{C}}\mathcal{M}), \end{cases}$$

where $T_{\mathbb{C}}\mathcal{M}$ denotes the complexified tangent bundle to $\mathcal{M}$.

Then there exists a sequence of non-negative real numbers $\{k_j\}_j$ and $\varepsilon > 0$ with the following significance. Whenever $k \in \mathbb{C} \setminus \{\pm k_j\}_j$ and $2 - \varepsilon < p < 2 + \varepsilon$, the boundary problem $(BVP_k)_l$ has a unique solution. Also, the solution satisfies

$$(14.2)\qquad \|\mathcal{N}(u)\|_{L^p(\partial\Omega)} + \|\mathcal{N}(\delta u)\|_{L^p(\partial\Omega)} \leq C\left(\|g\|_{L^p(\partial\Omega,\Lambda^l T_{\mathbb{C}}\mathcal{M})} + \|f\|_{L^p(\partial\Omega,\Lambda^{l+1}T_{\mathbb{C}}\mathcal{M})}\right)$$

for some positive constant C, independent of f, g.

Moreover, with k and p as above,

$$(14.3)\qquad \mathcal{N}(du) \in L^p(\partial\Omega) \Longleftrightarrow f \in L^{p,d}_{\rm nor}(\partial\Omega, \Lambda^{l+1}T_{\mathbb{C}}\mathcal{M})$$

and, if $f \in L^{p,d}_{\rm nor}(\partial\Omega, \Lambda^{l+1}T_{\mathbb{C}}\mathcal{M})$, there holds

$$(14.4)\qquad \|\mathcal{N}(du)\|_{L^p(\partial\Omega)} \leq C\left(\|f\|_{L^{p,d}_{\rm nor}(\partial\Omega,\Lambda^{l+1}T_{\mathbb{C}}\mathcal{M})} + \|g\|_{L^p(\partial\Omega,\Lambda^l T_{\mathbb{C}}\mathcal{M})}\right).$$

Finally, if k and p are as above, then

$$(14.5)\qquad \delta u = 0 \Longleftrightarrow g = 0.$$

A similar set of results is valid for the Hodge-dual problem of $(BVP_k)_l$.

Assuming this result for the time being, we are ready to discuss the Maxwell system in Ω. In the current setting, for each $l \in \{0, 1, ..., m\}$, this consists of determining $E \in C^0(\Omega, \Lambda^l T_{\mathbb{C}}\mathcal{M})$, the electric form, and $H \in C^0(\Omega, \Lambda^{l+1} T_{\mathbb{C}}\mathcal{M})$, the magnetic form, so that

$$(14.6) \qquad (Max_k)_l \begin{cases} dE - ikH = 0 \text{ in } \Omega, \\ \delta H + ikE = 0 \text{ in } \Omega, \\ \delta E = 0, \;\; dH = 0 \text{ in } \Omega, \\ \mathcal{N}(E), \mathcal{N}(H) \in L^p(\partial\Omega), \\ \nu \wedge E\big|_{\partial\Omega} = f \in L^{p,d}_{\rm nor}(\partial\Omega, \Lambda^{l+1} T_{\mathbb{C}}\mathcal{M}), \end{cases}$$

where $k \in \mathbb{C}$ is the so-called wave-number.

THEOREM 14.2. *Let the Riemannian manifold $\mathcal{M}$ be as in § 5 and let $\Omega \subset \mathcal{M}$ be an arbitrary Lipschitz domain. Also, fix $l \in \{0, 1, ..., m\}$.*

Then there exists a sequence of non-negative real numbers $\{k_j\}_j$ and $\varepsilon > 0$ so that, whenever $k \in \mathbb{C} \setminus \{\pm k_j\}_j$ and $2 - \varepsilon < p < 2 + \varepsilon$, the boundary problem $(Max_k)_l$ has a unique solution. The solution satisfies

$$(14.7) \qquad \|\mathcal{N}(E)\|_{L^p(\partial\Omega)} + \|\mathcal{N}(H)\|_{L^p(\partial\Omega)} \leq C\|f\|_{L^{p,d}_{\rm nor}(\partial\Omega, \Lambda^{l+1} T_{\mathbb{C}}\mathcal{M})}$$

for some positive constant C, independent of f.

PROOF. Select $\{k_j\}_j$ and $\varepsilon > 0$ so that the conclusions of Theorem 14.1 are valid, then solve (14.1) with f as in (14.6) and $g = 0$. The important observation is that the pair $E := u$, $H := -ik^{-1}du$ becomes a solution for (14.6). The estimate (14.7) is also seen from this. To justify uniqueness, simply note that if (E, H) is a null-solution for $(Max_k)_l$ then $u := E$ solves the homogeneous version of $(BVP_k)_l$. Hence, $u = 0$ which also entails $H = 0$. ■

We are therefore left with the

PROOF OF THEOREM 14.1. Consider the Hodge-Laplacian on l-forms, i.e.

$$(14.8) \qquad \Delta_l : H^{1,2}(\mathcal{M}, \Lambda^l T_{\mathbb{C}}\mathcal{M}) \to H^{-1,2}(\mathcal{M}, \Lambda^l T_{\mathbb{C}}\mathcal{M}).$$

Under the assumption that the Riemannian metric has $H^{2,r}$, $r > m$, coefficients, this is a bounded, negative, formally self-adjoint operator. In particular, $(\Delta_l - \lambda)^{-1}$ exists and is a self-adjoint, compact operator on $L^2(\mathcal{M}, \Lambda^l T_{\mathbb{C}}\mathcal{M})$ for $\lambda \in \mathbb{R}$ sufficiently large. This implies the existence of a discrete subset of $(-\infty, 0]$, which we denote by $\mathrm{Spec}\,(\Delta_l)$, which accumulates only at $-\infty$ and so that $\Delta_l - z$ is invertible from $H^{1,2}(\mathcal{M}, \Lambda^l T_{\mathbb{C}}\mathcal{M})$ onto $H^{-1,2}(\mathcal{M}, \Lambda^l T_{\mathbb{C}}\mathcal{M})$ whenever $z \notin \mathrm{Spec}\,(\Delta_l)$.

Let us select $k'_j \geq 0$ so that

$$(14.9) \qquad \{-(k'_j)^2\}_j := \bigcup_{0 \leq l \leq m} \mathrm{Spec}\,(\Delta_l).$$

Now, the entire discussion in § 6 can be carried out in connection with the operator $L := \Delta + k^2$ for any $k \in \mathbb{C} \setminus \{\pm k'_j\}_j$. To stress the dependence on k, we shall write M_k, N_k, $\mathcal{S}_k$, S_k, etc., for the resulting layer potential operators. These operators enjoy all the properties described in § 6 for the case of real-valued potentials V. One should also keep in mind that, since k is constant, the residual terms in (6.3) and (6.5) vanish.

Consider now $\varepsilon > 0$ so that the conclusion of Corollary 10.4 holds for some fixed, strictly positive constant potential V; for further reference, denote the boundary integral operators associated (as in (6.6), (6.9)) with this choice of the potential by M_V, N_V, S_V, etc. Next, for $k \in \mathbb{C} \setminus \{\pm k'_j\}_j$, write

$$\pm\tfrac{1}{2}I + N_k = (\pm\tfrac{1}{2}I + N_V)\big[I + T_{k,V}\big], \tag{14.10}$$

where $T_{k,V} := (\pm\frac{1}{2}I + N_V)^{-1}(N_k - N_V)$. It follows from the discussion in § 6 that $T_{k,V}$ is a compact operator both on $L^p_{\rm nor}(\partial\Omega, \Lambda^l T_{\mathbb{C}}\mathcal{M})$ and on $L^{p,d}_{\rm nor}(\partial\Omega, \Lambda^l T_{\mathbb{C}}\mathcal{M})$. Also, $T_{k,V}$ is the zero operator when $k = V$, and the mapping $k \mapsto T_{k,V}$ is holomorphic in $\mathbb{C}\setminus\{\pm k'_j\}_j$. Then the analytic Fredholm theorem implies that $I + T_{k,V}$ is invertible both on $L^p_{\rm nor}(\partial\Omega, \Lambda^l T_{\mathbb{C}}\mathcal{M})$ and on $L^{p,d}_{\rm nor}(\partial\Omega, \Lambda^l T_{\mathbb{C}}\mathcal{M})$, for all $k \in \mathbb{C}\setminus\{\pm k'_j\}_j$, except perhaps for a discrete subset $\{k''_j\}_j$ of $\mathbb{C} \setminus \{\pm k'_j\}_j$. By virtue of (14.10), the same conclusion applies to $\pm\frac{1}{2}I + N_k$.

The next remark is that $\pm\frac{1}{2}I + N_k$ is invertible (on the same two spaces) for any $k \in \mathbb{C} \setminus \mathbb{R}$. Indeed, Fredholmness with index zero follows from the fact that $N_k - N_V$ is compact, whereas injectivity is seen from certain energy identities as in [JM] (here, the regularity result described in Corollary 10.5 is also used); we omit the details.

Combining all these observations, we have proved that there exist $\varepsilon > 0$ and a sequence $\{k_j\}_j (= \{k'_j\}_j \cup \{|k''_j|\}_j)$ of real, positive numbers so that, for any $l \in \{0, 1, ..., m\}$,

$$\begin{gathered}\pm\tfrac{1}{2}I + N_k \text{ is invertible on } L^p_{\rm nor}(\partial\Omega, \Lambda^l T_{\mathbb{C}}\mathcal{M}) \text{ and on } L^{p,d}_{\rm nor}(\partial\Omega, \Lambda^l T_{\mathbb{C}}\mathcal{M})\\ \text{for any } k \in \mathbb{C} \setminus \{\pm k_j\}_j \text{ and any } 2 - \varepsilon < p < 2 + \varepsilon.\end{gathered} \tag{14.11}$$

Via an application of the Hodge $*$-isomorphism, this also gives that, with $\varepsilon > 0$ and $\{k_j\}_j$ as above,

$$\begin{gathered}\pm\tfrac{1}{2}I + M_k \text{ is invertible on } L^p_{\rm tan}(\partial\Omega, \Lambda^l T_{\mathbb{C}}\mathcal{M}) \text{ and on } L^{p,\delta}_{\rm tan}(\partial\Omega, \Lambda^l T_{\mathbb{C}}\mathcal{M})\\ \text{for any } k \in \mathbb{C} \setminus \{\pm k_j\}_j \text{ and any } 2 - \varepsilon < p < 2 + \varepsilon,\end{gathered} \tag{14.12}$$

for each $l \in \{0, 1, ..., m\}$.

Armed with these invertibility results, we can now tackle the first part in the conclusion of Theorem 14.1. Regarding existence, with $k \in \mathbb{C}\setminus\{\pm k_j\}_j$ and $2 - \varepsilon < p < 2 + \varepsilon$,

$$\begin{aligned}u :=& \delta\mathcal{S}_k\big[(\tfrac{1}{2}I + N_k)^{-1}(f - S_k(\tfrac{1}{2}I + N_k)^{-1}g))\big]\\ &+ \mathcal{S}_k\big[(\tfrac{1}{2}I + N_k)^{-1}g)\big], \qquad \text{in } \Omega,\end{aligned} \tag{14.13}$$

clearly solves $(BVP_k)_l$ and satisfies the estimate (14.2).

In order to show uniqueness, assume that u solves the homogeneous version of $(BVP_k)_l$. The first claim we make is that the same type of arguments used to prove Lemma 12.3 give that, in our case, $\mathcal{N}(du) \in L^p(\partial\Omega)$. The point is that an integral representation formula similar to (13.11) is valid with $V = -k^2$, since, in the case under discussion, $k^2 \notin \mathrm{Spec}\,(\Delta_l)$. Note that, this time all layer potentials are based on the Schwartz kernels of $(\Delta_l + k^2)^{-1}$, $0 \le l \le m$. Since, in this situation, there are no volume potentials involved, the claim follows from this.

Armed with this regularity result, it is not too difficult to prove a Green type integral representation formula for u, to the effect that

$$u = \delta\mathcal{S}_k(\nu \wedge u) - d\mathcal{S}_k(\nu \vee u) - \mathcal{S}_k(\nu \vee du) + \mathcal{S}_k(\nu \wedge \delta u) \text{ in } \Omega; \tag{14.14}$$

this is much in the spirit of (11.18). Thanks to the homogeneous boundary conditions in $(BVP_k)_l$, the first and last terms in the right side of (14.14) can be dropped. If, in the resulting identity, we now apply d to both sides, go nontangentially to the boundary and then take $\nu \vee \cdot$ of both sides, we end up with $(\frac{1}{2}I + M_k)(\nu \vee du) = 0$ on $\partial\Omega$. Because of (14.12), this further implies $\nu \vee du = 0$ on $\partial\Omega$. Using this back in (14.14) leaves us with $u = -dS_k(\nu \vee u)$ in Ω. Once again, going nontangentially to the boundary and taking $\nu \vee \cdot$ of both sides yields $(\frac{1}{2}I + M_k)(\nu \vee u) = 0$ on $\partial\Omega$ so that, by (14.12), $\nu \vee u = 0$ on $\partial\Omega$ also. Now $u = 0$ follows from (14.14). This concludes the proof of uniqueness.

The first regularity statement, i.e. (14.3), together with the estimate (14.4) follow directly from the explicit integral representation (14.13) of the solution and (14.11).

Finally, to see (14.5), if u is a solution for $(BVP_k)_l$ with $g = 0$ then $v := \delta u$ solves the homogeneous version of $(BVP_k)_{l-1}$. Hence, from what we have proved so far, v must vanish, i.e. $\delta u = 0$. The proof of Theorem 14.1 is therefore finished. ■

In closing, let us point out that appropriate versions of (14.11)-(14.12) are valid for the operators $\lambda I + N_k$, $\lambda I + M_k$, for $\lambda \in \mathbb{R}$, $|\lambda| \geq \frac{1}{2}$. Such results are useful when dealing with certain transmission problems for the Maxwell system. See [MM2] for the Euclidean case.

Appendix A

Analysis on Lipschitz Manifolds

Setting the stage for Appendix B, for the convenience of the reader, here we gather several useful rudiments of a function theory on Lipschitz manifolds.

A *topological manifold* Σ of dimension m is a Hausdorff, paracompact, second countable, topological space Σ with the property that for every $x \in \Sigma$ there exists an open set U in Σ, $x \in U$, and a mapping $\phi : U \to \mathbb{R}^m$ such that $\phi(U)$ is open in $\mathbb{R}^m$ and $\phi : U \to \phi(U)$ is a homeomorphism. We shall call (U, ϕ) a *coordinate chart* (about x). An *atlas* on Σ is a family $\mathcal{A} = \{U_i, \phi_i\}_{i \in I}$ such that $\Sigma = \bigcup_{i \in I} U_i$ and (U_i, ϕ_i) is a coordinate chart for each $i \in I$. A topological manifold Σ is called a *Lipschitz manifold* if there exists an atlas (called Lipschitz atlas) $\mathcal{A} = \{U_i, \phi_i\}_{i \in I}$ such that for any $i, j \in I$ the transition map $\phi_i \circ \phi_j^{-1} : \phi_j(U_i \cap U_j) \to \phi_i(U_i \cap U_j)$ is locally by-Lipschitz (with respect to the usual metric in $\mathbb{R}^m$).

Two atlases are called equivalent provided their union is an atlas. A *Lipschitz structure* on Σ is the equivalence class of a certain Lipschitz atlas. In what follows we shall fix a Lipschitz structure on Σ induced by a fixed Lipschitz atlas $\mathcal{A} = \{U_i, \phi_i\}_{i \in I}$.

Suppose that $(\Sigma_j, \mathcal{A}_j)$ are two Lipschitz manifolds, $j = 1, 2$. A mapping $f : \Sigma_1 \to \Sigma_2$ will be called *locally Lipschitz* if it is continuous and for any two coordinate charts $(U_j, \phi_j) \in \mathcal{A}_j$, $j = 1, 2$, the composition $\phi_2 \circ f \circ \phi_1^{-1} : \phi_1(U_1 \cap f^{-1}(U_2)) \to \phi_2(U_2)$ is a locally Lipschitz function.

A set $S \subseteq \Sigma$ is said to have *zero measure* if and only if $\phi_i(U_i \cap S)$ has measure zero in $\mathbb{R}^m$ with respect to the usual m-dimensional Lebesgue measure for any $(U_i, \phi_i) \in \mathcal{A}$. Accordingly, a property is said to hold *almost everywhere* on Σ provided it fails at most on a set of measure zero on Σ.

A real valued function defined a.e. on Σ is called *measurable* if it is so in any coordinate chart of the structural atlas. Furthermore, the class $L^p_{loc}(\Sigma)$, of real valued functions locally L^p-integrable on Σ, $1 \leq p \leq \infty$, is introduced in a similar fashion.

If Σ is also compact, then "locally" can be dropped and one may talk of (globally) *Lipschitz* functions on Σ, or $L^p(\Sigma)$, etc. Furthermore, we may select a structural atlas having only finitely many coordinate charts.

Next we introduce the *singular set* of Σ (corresponding to $\mathcal{A}$) as being

$$\begin{aligned} Sing(\Sigma) := \{x \in \Sigma;\ &\text{there exist } i, j \in I \text{ with } x \in U_i \cap U_j \text{ and such that the} \\ &\text{function } \phi_i \circ \phi_j^{-1} : \phi_j(U_i \cap U_j) \to \phi_i(U_i \cap U_j) \text{ is not differentiable at } \phi_j(x)\}. \end{aligned}$$

A basic observation is that the singular set of a Lipschitz manifold has measure zero. In the sequel, points in $Sing(\Sigma)$ will be called *singular points* (relative to $\mathcal{A}$), whereas points in $Reg(\Sigma) := \Sigma \setminus Sing(\Sigma)$ will be referred to as *regular points* (relative to $\mathcal{A}$).

Let now $(\Sigma_j, \mathcal{A}_j)$ be two Lipschitz manifolds, $j = 1, 2$. A mapping $f : \Sigma_1 \to \Sigma_2$ will be called *differentiable* at $x \in \Sigma_1$ provided the following are valid:

1. x is a regular point of Σ_1,
2. $f(x)$ is a regular point of Σ_2,
3. there exist $(U_j, \phi_j) \in \mathcal{A}_j$, $j = 1, 2$, with $x \in U_1$, $f(x) \in U_2$, such that the function $\phi_2 \circ f \circ \phi_1^{-1} : \phi_1(U_1 \cap f^{-1}(U_2)) \to \phi_2(U_2)$ is differentiable at $\phi_1(x)$.

If $x \in \Sigma$ two mappings f, g from a neighborhood of x into $\mathbb{R}$ are called *equivalent at* x (and we denote this by $f \overset{x}{\sim} g$) if there exists V open small neighborhood of x such that $f|_V = g|_V$. Classes of equivalence modulo $\overset{x}{\sim}$ will be called *germs at* x. We shall pay special attention to germs of locally Lipschitz functions that are differentiable at a regular point x, which will be denoted by $Lip^x_{diff}(\Sigma)$.

Let $x \in \Sigma$ be a regular point. A locally Lipschitz mapping $\gamma : (-\varepsilon, \varepsilon) \to \Sigma$, $\varepsilon > 0$, with $\gamma(0) = x$, γ differentiable at 0 will be called *path* (through x). For such a path γ we define a linear mapping $\frac{d}{d\gamma} : Lip^x_{diff}(\Sigma) \to \mathbb{R}$ called *derivation along* γ (at x) by $\frac{d}{d\gamma}([f]) \overset{def}{=} \frac{d}{dt}(f \circ \gamma)(t)\big|_{t=0}$, for any $[f] \in Lip^x_{diff}(\Sigma)$. If $(U, \phi) \in \mathcal{A}$ then the derivation along $\phi^{-1}(\phi(x) + te_i)$ at $x \in Reg(\Sigma)$ is denoted by $\frac{d}{d\phi_i}$, $i = 1, 2, ..., m$. Note that

$$\frac{d}{d\phi_i}([f]) = \frac{\partial(f \circ \phi^{-1})}{\partial x_i}(\phi(x)), \; i = 1, 2, ..., m. \tag{A.1}$$

The tangent space at $x \in \Sigma$ to the manifold Σ is defined by

$$T_x\Sigma := \left\{ \frac{d}{d\gamma}; \gamma \text{ path through } x \right\} \quad \text{if } x \text{ is regular}$$

and

$$T_x\Sigma := 0 \quad \text{if } x \text{ is singular}.$$

It is not difficult to check that $T_x\Sigma$ is a vector space and that in fact $dim(T_x\Sigma) = m$ (i.e. the same as the dimension of Σ) at any regular point x. In fact, for $(U, \phi) \in \mathcal{A}$ a basis in $T_x\Sigma$ at any regular point $x \in U$ is given by $\{\frac{d}{d\phi_i}\}_{i=1}^m$. Now, the *tangent bundle* is $T\Sigma := \bigsqcup_{x \in \Sigma} T_x\Sigma$.

Let $f : \Sigma_1 \to \Sigma_2$ be a locally Lipschitz function between two Lipschitz manifolds Σ_1 and Σ_2. The *differential* of f is the mapping $df : T\Sigma_1 \to T\Sigma_2$ defined almost everywhere in the following way. For $x \in Reg(\Sigma_1)$ such that $f(x) \in Reg(\Sigma_2)$, df_x maps $T_x\Sigma_1$ into $T_{f(x)}\Sigma_2$ by

$$df_x\left(\frac{d}{d\gamma}\right) := \frac{d}{d(f \circ \gamma)},$$

for any path γ through x (note that $f \circ \gamma$ is a path through $f(x)$). Also, if $(U, \phi) \in \mathcal{A}$ then $d\phi_j\left(\frac{d}{d\phi_k}\right) = \delta_{jk}\frac{d}{dt}$, where we have denoted by $\frac{d}{dt}$ the standard derivation on $\mathbb{R}$.

We are now in a position to introduce the class of differential forms with measurable coefficients on a Lipschitz manifold. To this end, let Σ be a Lipschitz manifold of dimension m, and let $\mathcal{A}$ be a structural atlas for Σ. For each $l = 0, 1, ..., m$, we let $\Lambda^l T\Sigma$ stand for the collection of all applications ω defined almost everywhere on Σ and taking values in $\Lambda^l T\Sigma := \bigsqcup_{x \in \Sigma} \Lambda^l T_x\Sigma$ satisfying

1. $\omega(x) \in \Lambda^l T_x\Sigma$ for a.e. $x \in \Sigma$;

2. for any $(U, \phi) \in \mathcal{A}$ and any multi-index J of length l there exist real valued, measurable functions a_J defined on U such that

$$\omega|_U = \sum_{|J|=l} a_J \, d\phi^J \quad \text{a.e. in } U. \tag{A.2}$$

This concept is well defined on Lipschitz manifolds since changing coordinates involves only multiplications with essentially bounded functions (arising as determinants of minors of the Jacobians of the transition functions).

Basically for the same reasons it makes sense to talk about l-forms with coefficients in $L^p_{loc}(\Sigma)$, for $1 \leq p \leq \infty$. That is, the functions a_J in the local representation formula (A.2) belong to $L^p_{loc}(\Sigma)$. We denote the set of such forms by $L^p_{loc}(\Sigma, \Lambda^l T\Sigma)$.

Next we define the pull-back operation. Once again let $f : \Sigma_1 \to \Sigma_2$ be a locally Lipschitz function between two Lipschitz manifolds. Then, $f^* : \Lambda^l T\Sigma_2 \to \Lambda^l T\Sigma_1$ is defined by

$$(f^*\omega)(x)\left(\frac{d}{d\gamma_1}, \cdots, \frac{d}{d\gamma_l}\right) := \omega(f(x))\left(df_x\left(\frac{d}{d\gamma_1}\right), \cdots, df_x\left(\frac{d}{d\gamma_l}\right)\right), \tag{A.3}$$

for any $\omega \in \Lambda^l T\Sigma_2$, any γ_j, $j = 1, 2, .., l$, paths through x at almost every $x \in \Sigma_1$. It is not too difficult to check that this definition is meaningful and that the variant of the pull-back introduced here enjoys basically all the functorial properties of the classical one.

Let us point out that $\Lambda^l T\Sigma$ can be naturally identified with the collection of families $\{\omega_U\}_{(U,\phi)\in\mathcal{A}}$, where each $\omega_U : \phi(U) \to \Lambda^l \mathbb{R}^m$ is measurable and the following compatibility condition is satisfied

$$(\phi_i \circ \phi_j^{-1})^* \omega_{U_j}\big|_{U_i \cap U_j} = \omega_{U_i}\big|_{U_i \cap U_j}, \quad \text{for any } i, j \in I.$$

This identification is particularly useful for defining the *exterior derivative operator* d_Σ on a Lipschitz manifold Σ. More concretely, if $\omega \in L^1_{loc}(\Sigma, \Lambda^l T\Sigma)$ is such that, in the distributional sense $d_{\mathbb{R}^m}\omega_U \in L^1_{loc}(\phi(U), \Lambda^{l+1}\mathbb{R}^m)$ for all $(U, \phi) \in \mathcal{A}$, then we set

$$d_\Sigma \omega := \{d_{\mathbb{R}^m}\omega_U\}_{(U,\phi)\in\mathcal{A}}$$

(here m =the dimension of Σ). Once again all important properties of this operator have suitable extensions in this setting. In particular, $d_\Sigma d_\Sigma = 0$, and d_Σ commutes with the pull-back operator, i.e.

$$f^* \, d_{\Sigma_2} = d_{\Sigma_1} \, f^*, \tag{A.4}$$

if f is a locally Lipschitz mapping between two Lipschitz manifolds Σ_1, Σ_2.

Assume now that the Lipschitz manifold Σ is *oriented* and also equipped with a *(Lipschitz) Riemannian metric.* Being oriented is defined essentially as in the smooth case. That is, an orientation has been specified in $T_x\Sigma$ for a.e. $x \in \Sigma$ such that there exists a structural atlas $\mathcal{A}$ which contains only positive coordinate charts. A chart $(U, \phi) \in \mathcal{A}$ is called *positive* if $\left\{\frac{d}{d\phi_1}, ..., \frac{d}{d\phi_m}\right\}$ is a positively oriented basis of $T_x\Sigma$ for a.e. $x \in U$.

Also by a *Lipschitz Riemannian structure*, we mean that at almost any point $x \in \Sigma$ some inner product $\langle \cdot, \cdot \rangle_x$ has been specified on the tangent space $T_x\Sigma$ with the following properties:

1. $\langle \cdot, \cdot \rangle_x$ varies measurably with x, that is if $(U, \phi) \in \mathcal{A}$ then the functions
$$g_{ij}^U(x) := \langle \frac{d}{d\phi_i}, \frac{d}{d\phi_j} \rangle_x$$
are measurable on U;
2. there exists a structural atlas $\mathcal{A}$ and two constants $C_1, C_2 > 0$ such that for any $(U, \phi) \in \mathcal{A}$, for a.e. $x \in U$, and any path γ through x we have

$$C_1 \|(\phi \circ \gamma)'(0)\|_{\mathbb{R}^m}^2 \leq \langle \frac{d}{d\gamma}, \frac{d}{d\gamma} \rangle_x \leq C_2 \|(\phi \circ \gamma)'(0)\|_{\mathbb{R}^m}^2 \tag{A.5}$$

(here, the norm $\|\cdot\|_{\mathbb{R}^m}$ refers to the Euclidean norm in $\mathbb{R}^m$).

This latter condition implies that the matrix $G^U(x) := (g_{i,j}^U(x))_{1 \leq i,j \leq m}$ is symmetric, bounded and positive definite uniformly in $x \in U$. In fact,

$$C_1 \|v\|_{\mathbb{R}^m}^2 \leq \langle G^U(x)v, v \rangle_{\mathbb{R}^m} \leq C_2 \|v\|_{\mathbb{R}^m}^2, \tag{A.6}$$

for any $v \in \mathbb{R}^m$ and a.e. $x \in U$. In fact, any Lipschitz manifold has a Lipschitz Riemannian metric (this can be obtained by transferring the Euclidean metric from $\mathbb{R}^m$ in the usual fashion).

In the usual fashion, the inner product on the tangent space $T_x\Sigma$ induces a natural inner product on $\Lambda^l T_x\Sigma$ for each $0 \leq l \leq m$ at a.e. $x \in \Sigma$. This will be denoted by $\langle \cdot, \cdot \rangle_{\Lambda^l T\Sigma}$. In particular, the unique form ω of maximal degree, normalized such that $\langle \omega, \omega \rangle_{\Lambda^m \Sigma} = 1$ a.e. on Σ and which is positively oriented, is called the volume element on Σ and will be denoted by dV_Σ. As regards integration on Σ, if $f \in L^1_{comp}(\Sigma)$ then

$$\int_\Sigma f \, dV_\Sigma = \sum_j \int_{\phi_j(U_j)} (\phi_j^{-1})^*(\theta_j f dV_\Sigma), \tag{A.7}$$

where $\{\theta_j\}_j$ is a Lipschitz partition of unity on Σ subordinated to (a locally finite) open cover $(U_j)_j$ of Σ, $(U_j, \phi_j) \in \mathcal{A}$. It is not difficult to see that for any $(U, \phi) \in \mathcal{A}$,

$$(\phi^{-1})^*(dV_\Sigma) = \left[det \left(\langle \frac{d}{d\phi_i}, \frac{d}{d\phi_j} \rangle_{\phi^{-1}(\cdot)} \right)_{i,j} \right]^{1/2} dV_{\mathbb{R}^m}. \tag{A.8}$$

Observe that from (A.5) it follows that if $(U, \phi) \in \mathcal{A}$ then

$$C_1^{m/2} \leq \left[det \left(\langle \frac{d}{d\phi_i}, \frac{d}{d\phi_j} \rangle_x \right)_{i,j} \right] \leq C_2^{m/2} \tag{A.9}$$

at a.e. $x \in U$.

In the case in which Σ is compact we note that, for each $1 \leq p < \infty$, $L^p(\Sigma, \Lambda^l T\Sigma)$ becomes a Banach space when equipped with the norm

$$\|\omega\|_{L^p(\Sigma, \Lambda^l T\Sigma)} := \left(\int_\Sigma \langle \omega, \omega \rangle_{\Lambda^l T\Sigma}^{p/2} \, dV_\Sigma \right)^{1/p} \tag{A.10}$$

(for $p = 2$ this is in fact a Hilbert space). Observe that, two different Lipschitz Riemannian metrics define equivalent norms on $L^p(\Sigma, \Lambda^l T\Sigma)$.

Let Σ_1, Σ_2 be compact, oriented Riemannian Lipschitz manifolds $f : \Sigma_1 \to \Sigma_2$ a Lipschitz mapping and $1 \leq p < \infty$. Given (A.6)-(A.9), it is not difficult to see that there exists a positive constant C depending only on the Lipschitz character of Σ_1, Σ_2 and f, such that for any $0 \leq l \leq dim\Sigma_2$ and any $\omega \in L^p(\Sigma_2, \Lambda^l T\Sigma_2)$ we have that $f^*\omega \in L^p(\Sigma_1, \Lambda^l \Sigma_1)$ and

$$\|f^*\omega\|_{L^p(\Sigma_1,\Lambda^l T\Sigma_1)} \leq C\|\omega\|_{L^p(\Sigma_2,\Lambda^l T\Sigma_2)}.$$

In other words, $f^* : L^p(\Sigma_2, \Lambda^l T\Sigma_2) \to L^p(\Sigma_1, \Lambda^l T\Sigma_1)$ is well-defined, linear and bounded.

Appendix B

The Connection Between d_∂ and $d_{\partial\Omega}$

Let $\mathcal{M}$ be a Riemannian manifold (as in §1) and $\Omega \subset \mathcal{M}$ a Lipschitz subdomain. When equipped with the metric inherited from $\mathcal{M}$, $\partial\Omega$ becomes a Lipschitz manifold endowed with a Lipschitz Riemannian metric.

In this appendix we make explicit the relationship between our boundary derivative operator d_∂ (see (5.14)) and the intrinsic exterior derivative operator $d_{\partial\Omega}$ corresponding to the Lipschitz manifold $\partial\Omega$ (cf. Appendix A). This investigation is particularly important in the calculation of the cohomology groups associated with the boundary complex (9.11); cf. (9.14) and the proof of Lemma 9.4 at the end of this appendix.

Denote by Meas$\,(\partial\Omega, \Lambda^l T\mathcal{M})$ the vector space of (global) measurable sections in the bundle $\Lambda^l T\mathcal{M}|_{\partial\Omega}$ over the Lipschitz manifold $\partial\Omega$. Similarly, Meas$\,(\partial\Omega, \Lambda^l T\partial\Omega)$ stands for the space of global, measurable sections in the vector bundle of l-forms on the Lipschitz manifold $\partial\Omega$. The latter makes sense according to the discussion in Appendix A. For each such $x \in \partial\Omega$, we set π_x for the orthogonal projection of $T_x\mathcal{M}$ onto $T_x\partial\Omega$ and introduce

$$\pi_\star : \text{Meas}\,(\partial\Omega, \Lambda^l T\partial\Omega) \to \text{Meas}\,(\partial\Omega, \Lambda^l T\mathcal{M}) \tag{B.1}$$

defined by

$$(\pi_\star\omega)(x)(X_1, ..., X_l) := \omega(x)(\pi_x X_1, ..., \pi_x X_l) \tag{B.2}$$

for each $\omega \in \text{Meas}\,(\partial\Omega, \Lambda^l T\partial\Omega)$, a.e. $x \in \partial\Omega$ and each $X_1, ..., X_l \in T_x\mathcal{M}$. We shall also work with a "lifting" operator,

$$j^\star : \text{Meas}\,(\partial\Omega, \Lambda^l T\mathcal{M}) \to \text{Meas}\,(\partial\Omega, \Lambda^l T\partial\Omega) \tag{B.3}$$

defined at a.e. $x \in \partial\Omega$ by

$$(j^\star w)(x)(X_1, ..., X_l) := w(x)(\iota(X_1), .., \iota(X_l)) \tag{B.4}$$

where $w \in \text{Meas}\,(\partial\Omega, \Lambda^l T\mathcal{M})$, $X_1, ..., X_l \in T_x\partial\Omega$, and ι is the inclusion of $T_x\partial\Omega$ into $T_x\mathcal{M}$. The main elementary properties of these operators are summarized in the next lemma.

LEMMA B.1. *For each $0 \le l \le m$ the following hold:*
1. *If $\omega \in Meas\,(\partial\Omega, \Lambda^l T\partial\Omega)$ then $j^\star\pi_\star\omega = \omega$.*
2. *If $\omega \in Meas\,(\partial\Omega, \Lambda^l T\mathcal{M})$ then $\omega_{\tan} = \pi_\star j^\star\omega$.*
3. *$\omega \in Meas\,(\partial\Omega, \Lambda^l T\mathcal{M})$ is normal if and only if $j^\star\omega = 0$.*
4. *$\pi_\star\omega$ is tangential for any $\omega \in Meas\,(\partial\Omega, \Lambda^l T\partial\Omega)$.*
5. *For any $u \in Meas\,(\partial\Omega, \Lambda^l T\mathcal{M})$ and any $v \in Meas\,(\partial\Omega, \Lambda^l T\partial\Omega)$ we have*

$$\langle j^\star u\,,\, v\rangle_{\Lambda^l T\partial\Omega} = \langle u,\, \pi_\star v\rangle_{\Lambda^l T\mathcal{M}} \ \ a.e.\ on\ \ \partial\Omega. \tag{B.5}$$

6. *If ω is a continuous (global) section in $\Lambda^l T\mathcal{M}$, $w \in Meas(\partial\Omega, \Lambda^l T\partial\Omega)$, and if ι is the inclusion of $\partial\Omega$ into $\mathcal{M}$ then $\iota^*\omega = j^\star(\omega|_{\partial\Omega})$ and*

$$\langle \iota^*\omega\,,\, w\rangle_{\Lambda^l T\partial\Omega} = \langle \omega|_{\partial\Omega},\, \pi_\star w\rangle_{\Lambda^l T\mathcal{M}} \text{ a.e. on } \partial\Omega. \tag{B.6}$$

Also,

$$|\iota^*\omega|_{\Lambda^l T\partial\Omega} = |(\omega|_{\partial\Omega})_{\tan}|_{\Lambda^l T\mathcal{M}} \text{ a.e. on } \partial\Omega. \tag{B.7}$$

The proof of this lemma is left to the interested reader. Based on this, it is trivial to establish the following.

COROLLARY B.2. *For $1 < p < \infty$ and $0 \le l \le m$, the mappings*

$$j^\star\,:\, L^p_{\tan}(\partial\Omega, \Lambda^l T\mathcal{M}) \to L^p(\partial\Omega, \Lambda^l \partial\Omega)$$

and

$$\pi_\star\,:\, L^p(\partial\Omega, \Lambda^l \partial\Omega) \to L^p_{\tan}(\partial\Omega, \Lambda^l T\mathcal{M})$$

are isometries inverse to each other.

Let $d_{\partial\Omega}$ denote the intrinsic exterior derivative operator on the Lipschitz manifold $\partial\Omega$. For each $1 < p < \infty$ and $0 \le l \le m$ we consider the maximal closed (unbounded) operator defined by $d_{\partial\Omega}$ on $L^p(\partial\Omega, \Lambda^l T\partial\Omega)$. Its domain is

$$\mathrm{Dom}_{l,p}(d_{\partial\Omega}) := \{\omega \in L^p(\partial\Omega, \Lambda^l T\partial\Omega);\ d_{\partial\Omega}\omega \in L^p(\partial\Omega, \Lambda^{l+1} T\partial\Omega)\}. \tag{B.8}$$

We shall also work with $\delta_{\partial\Omega} := (d_{\partial\Omega})^*$, the (functional analytic) adjoint of (the unbounded operator) $d_{\partial\Omega}$ on $L^p(\partial\Omega, \Lambda^l T\partial\Omega)$, whose domain we denote by $\mathrm{Dom}_{l,q}(\delta_{\partial\Omega})$ ($\subset L^q(\partial\Omega, \Lambda^l T\partial\Omega)$, $1/p + 1/q = 1$). Whenever irrelevant or clear from the context, we will drop the dependence of the domain on the integrability exponent.

LEMMA B.3. *For each $1 < p < \infty$ and $0 \le l \le m$ there hold*

$$\delta_\partial = \pi_\star\, \delta_{\partial\Omega}\, j^\star \tag{B.9}$$

and

$$\delta_{\partial\Omega} = j^\star\, \delta_\partial\, \pi_\star. \tag{B.10}$$

Here $j^\star$ and $\pi_\star$ are as in Corollary B.2 and all compositions are considered in the sense of compositions of unbounded operators.

PROOF. Let us deal with (B.9); the proof of (B.10) is fairly similar and is left to the reader. We will prove that if $f \in L^{p,\delta}_{\tan}(\partial\Omega, \Lambda^l T\mathcal{M})$ then $j^\star f \in \mathrm{Dom}_l(\delta_{\partial\Omega})$ and $\delta_{\partial\Omega}(j^\star f) = j^\star(\delta_\partial f)$. For such an f, we know from Corollary B.2 that $j^\star f \in L^p(\partial\Omega, \Lambda^l T\partial\Omega)$ and $j^\star(\delta_\partial f) \in L^p(\partial\Omega, \Lambda^{l-1} T\partial\Omega)$. Next, fix $\omega \in \mathrm{Dom}_{l-1}(d_{\partial\Omega})$. Using a partition of unity, pulling back things to the Euclidean space and then using a standard mollifying argument, it is not difficult to see that there exists a sequence $\{\psi_i\}_i$ of smooth $(l-1)$-forms in $\mathcal{M}$ with $\iota^*\psi_i \to \omega$ in $L^q(\partial\Omega, \Lambda^{l-1}T\partial\Omega)$ and $d_{\partial\Omega}(\iota^*\psi_i) \to d_{\partial\Omega}\omega$ in $L^q(\partial\Omega, \Lambda^l T\partial\Omega)$, $1/p + 1/q = 1$. Then, we have:

$$\begin{aligned}
&\int_{\partial\Omega} \langle j^\star(\delta_\partial f)\,,\,\omega\rangle_{\Lambda^{l-1}\partial\Omega}\,d\sigma \\
&\quad = \int_{\partial\Omega} \langle \delta_\partial f\,,\,\pi_\star\omega\rangle_{\Lambda^{l-1}T\mathcal{M}}\,d\sigma && \text{(by (B.5))} \\
&\quad = \lim_{i\to\infty} \int_{\partial\Omega} \langle \delta_\partial f\,,\,\pi_\star\iota^*\psi_i\rangle_{\Lambda^{l-1}T\mathcal{M}}\,d\sigma && \text{(by Corollary B.2)} \\
&\quad = \lim_{i\to\infty} \int_{\partial\Omega} \langle \delta_\partial f\,,\,(\psi_i)_{\tan}\rangle_{\Lambda^{l-1}T\mathcal{M}}\,d\sigma && \text{((2) and (6) in Lemma B.1)} \\
&\quad = \lim_{i\to\infty} \int_{\partial\Omega} \langle \delta_\partial f\,,\,\psi_i\rangle_{\Lambda^{l-1}T\mathcal{M}}\,d\sigma && (\delta_\partial f \text{ is tangential}) \\
&\quad = \lim_{i\to\infty} \int_{\partial\Omega} \langle f\,,\,(d_{\mathcal{M}}\psi_i)|_{\partial\Omega}\rangle_{\Lambda^{l}T\mathcal{M}}\,d\sigma && (\text{definition of } \delta_\partial) \\
&\quad = \lim_{i\to\infty} \int_{\partial\Omega} \langle \pi_\star j^\star f\,,\,(d_{\mathcal{M}}\psi_i)|_{\partial\Omega}\rangle_{\Lambda^{l}T\mathcal{M}}\,d\sigma && \text{((2) and (3) in Lemma B.1)} \\
&\quad = \lim_{i\to\infty} \int_{\partial\Omega} \langle j^\star f\,,\,\iota^* d_{\mathcal{M}}\psi_i\rangle_{\Lambda^{l}T\partial\Omega}\,d\sigma && \text{((6) in Lemma B.1)} \\
&\quad = \lim_{i\to\infty} \int_{\partial\Omega} \langle j^\star f\,,\,d_{\partial\Omega}\iota^*\psi_i\rangle_{\Lambda^{l}T\partial\Omega}\,d\sigma && (\text{as } \iota^* d_{\mathcal{M}} = d_{\partial\Omega}\iota^*) \\
&\quad = \int_{\partial\Omega} \langle j^\star f\,,\,d_{\partial\Omega}\omega\rangle_{\Lambda^{l}T\partial\Omega}\,d\sigma.
\end{aligned}$$

Since ω is arbitrary in $\mathrm{Dom}_{l-1}(d_{\partial\Omega})$, we conclude that $j^\star f$ belongs to $\mathrm{Dom}_l(\delta_\partial)$ and $\delta_{\partial\Omega}(j^\star f) = j^\star(\delta_\partial f)$ as desired. Finally, by applying $\pi_\star$ to this identity and using (2) in Lemma B.1 we see that $\pi_\star\delta_{\partial\Omega}j^\star f = \pi_\star j^\star(\delta_\partial f) = \delta_\partial f$. Thus, the proof is complete. ∎

PROPOSITION B.4. *For $g \in L^{p,d}_{\mathrm{nor}}(\partial\Omega, \Lambda^l T\mathcal{M})$ we have $j^\star(\nu\vee g) \in \mathrm{Dom}_{l-1}(d_{\partial\Omega})$ and*

$$d_{\partial\Omega}(j^\star(\nu\vee g)) = -j^\star(\nu\vee d_\partial g). \tag{B.11}$$

Conversely, if $\omega \in \mathrm{Dom}_l(d_{\partial\Omega})$ then $\nu\wedge\pi_\star\omega \in L^{p,d}_{\mathrm{nor}}(\partial\Omega, \Lambda^{l+1}T\mathcal{M})$ and

$$d_\partial(\nu\wedge\pi_\star\omega) = -\nu\wedge\pi_\star d_{\partial\Omega}\,\omega. \tag{B.12}$$

In particular, we have that

$$d_\partial = -\nu\wedge\pi_\star d_{\partial\Omega}\,j^\star\nu\vee\cdot \quad \textit{and} \quad d_{\partial\Omega} = -j^\star\nu\vee d_\partial(\nu\wedge\pi_\star) \tag{B.13}$$

in the sense of composition of unbounded operators.

PROOF. Since $d_{\partial\Omega}$ is closed, it follows that $d_{\partial\Omega} = (\delta_{\partial\Omega})^*$ in the sense of unbounded operators. Hence, for a fixed $g \in L^{p,d}_{\mathrm{nor}}(\partial\Omega, \Lambda^l T\mathcal{M})$, we have $j^\star(\nu\vee g) \in \mathrm{Dom}_{l-1}(d_{\partial\Omega})$ if and only if $j^\star(\nu\vee g) \in \mathrm{Dom}_{l-1}((\delta_{\partial\Omega})^*)$. Further, the latter membership is true if (and only if) there exists a form $\omega \in L^q(\partial\Omega, \Lambda^l T\partial\Omega)$, $1/p + 1/q = 1$, such that

$$\int_{\partial\Omega} \langle \delta_{\partial\Omega}w\,,\,j^\star(\nu\vee g)\rangle_{\Lambda^{l-1}T\partial\Omega}\,d\sigma = \int_{\partial\Omega} \langle w, \omega\rangle_{\Lambda^l T\partial\Omega}\,d\sigma \tag{B.14}$$

for any $w \in \mathrm{Dom}_l(\delta_{\partial\Omega})$, and in this case $d_{\partial\Omega}(j^\star(\nu \vee g)) = \omega$. To see that this is the case, fix $w \in \mathrm{Dom}_l(\delta_{\partial\Omega})$. Then,

$$\begin{aligned}
&\int_{\partial\Omega} \langle \delta_{\partial\Omega} w \,,\, j^\star(\nu \vee g)\rangle_{\Lambda^{l-1}T\partial\Omega}\, d\sigma \\
&\quad = \int_{\partial\Omega} \langle \pi_\star \delta_{\partial\Omega} w \,,\, \nu \vee g\rangle_{\Lambda^{l-1}T\mathcal{M}}\, d\sigma \quad \text{(by (B.5))} \\
&\quad = \int_{\partial\Omega} \langle \delta_\partial \pi_\star w \,,\, \nu \vee g\rangle_{\Lambda^{l-1}T\mathcal{M}}\, d\sigma \quad \text{(by Lemma B.3)} \\
&\quad = -\int_{\partial\Omega} \langle \nu \wedge \pi_\star w,\, d_\partial g\rangle_{\Lambda^{l+1}T\mathcal{M}}\, d\sigma \quad \text{(by Proposition 9.1)} \\
&\quad = -\int_{\partial\Omega} \langle w,\, j^\star(\nu \vee d_\partial g)\rangle_{\Lambda^{l}T\partial\Omega}\, d\sigma \quad \text{(by (B.5))}.
\end{aligned}$$

Therefore, $j^\star(\nu \vee g) \in \mathrm{Dom}_{l-1}(d_{\partial\Omega})$ and $d_{\partial\Omega}(j^\star(\nu \vee g)) = -j^\star(\nu \vee d_\partial g)$, which is precisely (B.11). Also, by (2) in Lemma B.1 and the fact that $d_\partial g$ is normal, this identity implies further

$$\nu \wedge \pi_\star d_{\partial\Omega}\, j^\star \nu \vee g = -\nu \wedge (\pi_\star\, j^\star(\nu \vee d_\partial g)) = -\nu \wedge (\nu \vee d_\partial g) = -d_\partial g,$$

i.e., the first equality in (B.13).

The proof of (B.12) as well as the proof of the second identity in (B.13) are based on similar ideas and are left to the reader. ■

We are finally in a position to present the

PROOF OF LEMMA 9.4. The fact that $\ker d^1_\partial = \mathbb{R}\,\nu$ follows easily from the fact that

$$L^{p,d}_{\mathrm{nor}}(\partial\Omega, \Lambda^1 T\mathcal{M}) = \{f\nu;\ f \in H^{1,p}(\partial\Omega)\}, \quad 1 < p < \infty, \tag{B.15}$$

and the identity

$$|d_\partial(f\nu)| = |d_{\partial\Omega} f| \tag{B.16}$$

(which, in turn is seen from (B.13) and Lemma B.1).

Therefore, the only thing left to verify is the exactness of the complex (9.11) in the sense of sheaf theory. To this end, we note that by Proposition B.4, matters can be readily reduced to verifying a L^p-Poincaré type lemma ($1 < p < \infty$) for the operator $d_{\partial\Omega}$ locally on $\partial\Omega$. More concretely, we need to check that if $\omega \in \mathrm{Dom}_{l,p}(d_{\partial\Omega})$ has $d_{\partial\Omega}\omega = 0$ in a neighborhood of a boundary point $x_0 \in \partial\Omega$, then there exists $w \in \mathrm{Dom}_{l-1,p}(d_{\partial\Omega})$ such that $d_{\partial\Omega} w = \omega$ in a (possible smaller) neighborhood of x_0.

However, such a statement is invariant to pull-back and, therefore, can be transferred to a ball in $\mathbb{R}^{m-1}$ in which setting the result is essentially well-know; cf. also the proof of Theorem 11.1. ■

Bibliography

[Ar] N. Aronszajn, *A unique continuation theorem for solutions of elliptic partial differential equations or inequalities of second order*, Journ. de Math. **36** (1957), 235–249.

[AKS] N. Aronszajn, A. Krzywicki and J. Szarski, *A unique continuation theorem for exterior differential forms on Riemannian manifold*, Arkiv för Matematik **4** (1962), 417–453.

[Bi] A. V. Bitsadze, *On the uniqueness of the solution of the Dirichlet problem for elliptic PDE's*, Uspekhi Matematicheskikh Nauk (N.S.) **3** (1948), 211–212.

[Bo] J.-M. Bony, *Calcul symbolique et propagation des singularités pour les équa- tions aux derivées partielles nonlinéaires*, Ann. Sci. Ecole Norm. Sup. **14** (1981), 209–246.

[Bou] G. Bourdaud, *L^p-estimates for certain non regular pseudo-differential operators*, Comm. PDE **7** (1982), 1023–1033.

[CFK] L. A. Caffarelli, E. B. Fabes and C. E. Kenig, *Completely singular elliptic-harmonic measures*, Indiana Univ. Math. J. **30** (1981), 917–924.

[Ca1] A. Calderón, *Cauchy integrals on Lipschitz curves and related operators*, Proc. NAS USA **74** (1977), 1324–1327.

[Ca2] A. P. Calderón, *Commutators, singular integrals on Lipschitz curves and applications*, Proc. Int. Congr. Math., Helsinki, 1, 1980, pp. 85–96.

[Ca3] A. P. Calderón, *Boundary value problems for the Laplace equation in Lipschitzian domains*, Recent Progress in Fourier Analysis (I. Peral and J. Rubio de Francia, ed.), Elsevier/North-Holland, Amsterdam, 1985, pp. 33–48.

[CS] W. Cao and Y. Sagher, *Stability of Fredholm properties on interpolation scales*, Ark. Math. **28** (1990), 249–258.

[Ch] M. Christ, *Lectures on Singular Integral Operators*, vol. 77, CBMS Series in Math., 1990.

[CJ] M. Christ and J.-L. Journé, *Polynomial growth estimates for multilinear singular integral operators*, Acta Math. **4** (1988), 219–225.

[CMM] R. Coifman, A. McIntosh, and Y. Meyer, *L'intégrale de Cauchy definit un opérateur borné sur L^2 pour les courbes lipschitziennes*, Annals of Math. **116** (1982), 361–388.

[Co] P. E. Conner, *The Green's and Neumann's problems for differential forms on Riemannian manifolds*, Proc. Nat. Acad. Sci. **40** (1954), 1151–1155.

[Da] B. Dahlberg, *On estimates of harmonic measure*, Arch. Rat. Mech. Anal. **65** (1977), 275–288.

[DK] B. Dahlberg and C. Kenig, *Hardy spaces and the L^p–Neumann problem for Laplace's equation in a Lipschitz domain*, Annals of Math. **125** (1987), 437–465.

[DKV] B. Dahlberg, C. Kenig and G. Verchota, *Boundary value problems for the system of elastostatics on Lipschitz domains*, Duke Math. J. **57** (1988), 795–818.

[DKPV] B. Dahlberg, C. Kenig, J. Pipher and G. Verchota, *Area integral estimates for higher order elliptic equations and systems*, Ann. Inst. Fourier Grenoble **47** (1997), 1425–1461.

[DR1] G. de Rham, *Sur l'analysis situs des variétés à n dimensiones*, J. Math. Pures Appl. **10** (1931), 115–200.

[DR2] G. de Rham, *Differentiable Manifolds*, Springer-Verlag, 1984.

[DS] S. K. Donaldson and D. P. Sullivan, *Quasiconformal 4-manifolds*, Acta Math. **163** (1989), 181-252.

[Du1] G. F. D. Duff, *Differential forms in manifolds with boundary*, Annals of Math. **56** (1952), 115–127.

[Du2] G. F. D. Duff, *Boundary value problems associated with the tensor Laplace equation*, Canadian J. Math. **5** (1953), 196–210.

[Du3] G. F. D. Duff, *A tensor equation of elliptic type*, Canadian J. Math. **5** (1953), 524–535.

[Du4] G. F. D. Duff, *A tensor boundary value problem of mixed type*, Canadian J. Math. **6** (1954), 427–440.

[DS1] G. F. D. Duff and D. C. Spencer, *Harmonic tensors on manifolds with boundary*, Proc. Nat. Acad. Sci. USA **37** (1951), 614–619.

[DS2] G. F. D. Duff and D. C. Spencer, *Harmonic tensors on Riemannian manifolds with boundary*, Annals of Math. **56** (1952), 128–156.

[Fa] E. Fabes, *Layer potential methods for boundary value problems on Lipschitz domains*, Potential Theory, Surveys and Problems, Lecture Notes in Math. (J. Král et al., ed.), vol. 1344, Springer-Verlag, 1988, pp. 55–80..

[FJR] E. Fabes, M. Jodeit and N. Rivière, *Potential techniques for boundary value problems on C^1 domains*, Acta Math. **141** (1978), 165–186.

[FKV] E. Fabes, C. Kenig, and G. Verchota, *The Dirichlet problem for the Stokes system on Lipschitz domains*, Duke Math. J. **57** (1988), 769–793.

[Fr1] K. O. Friedrichs, *The identification of weak and strong extensions of differential operators*, Trans. of Amer. Math. Soc. **55** (1944), 132–151.

[Fr2] K. O. Friedrichs, *Differential forms on Riemannian manifolds*, Comm. Pure Appl. Math. **8** (1955), 551–590.

[Gaf] M. Gaffney, *The harmonic operator for exterior differential forms*, Proc. Nat. Acad. Sci. U. S. A. **37** (1951), 48–50.

[Ga] W. Gao, *Layer potentials and boundary value problems for elliptic systems in Lipschitz domains*, J. Funct. Anal. **95** (1991), 377–399.

[GS1] P. R. Garabedian and D. C. Spencer, *Complex boundary value problems*, Trans. of Amer. Math. Soc. **73** (1952), 223–242.

[GS2] P. R. Garabedian and D. C. Spencer, *A complex tensor calculus for Kähler manifolds*, Acta Math. **89** (1953), 279–331.

[Gi] M. Giaquinta, *Multiple Integrals in the Calculus of Variations and Nonlinear Elliptic Systems,*, Annals of Math. Studies, Princeton University Press, Princeton, New Jersey, 1983.

[Go] R. Godement, *Topologie Algébrique et Théorie des Faisceaus*, Hermann, Paris, 1958.

[Hod1] W. V. D. Hodge, *A Dirichlet problem for harmonic functionals with applications to analytic varieties*, Proc. of London Math. Soc., Series 2 **36** (1934), 257–303.

[Hod2] W. V. D. Hodge, *The Theory and Applications of Harmonic Integrals*, Cambridge University Press, 1941.

[Ho] S. Hofmann, *Parabolic singular integrals of Calderón-type, rough operators and caloric layer potentials*, Duke Math. J. **90** (1997), 209–260.

[Hö] L. Hörmander, *L^2-Estimates and existence theorems for the $\bar{\partial}$-operator*, Acta Math. **113** (1965), 89–152.

[IM] T. Iwaniec, G. Martin, *Quasiregular mappings in even dimensions*, Acta Math. **170** (1993), 29-81.

[IMS] T. Iwaniec, M. Mitrea and C. Scott, *Boundary estimates for harmonic forms*, Proc. of Amer. Math. Soc. **124** (1996), 1467-1471.

[JM] B. Jawerth and M. Mitrea, *Higher dimensional scattering theory on C^1 and Lipschitz domains*, American Journal of Mathematics **117** (1995), 929–963.

[JK1] D. Jerison and C. Kenig, *The Neumann problem on Lipschitz domains*, Bull. Amer. Math. Soc. **4** (1981), 203–207.

[JK2] D. Jerison and C. Kenig, *The Dirichlet problem in non-smooth domains*, Annals of Math. **113** (1981), 367–382.

[JK3] D. Jerison and C. Kenig, *The inhomogeneous Dirichlet problem in Lipschitz domains*, J. Funct. Anal. **130** (1995), 161–219.

[KM] N. Kalton and M. Mitrea, *Stability results on interpolation scales of quasi-Banach spaces and applications*, Trans. Amer. Math. Soc. **350** (1998), 3903–3922.

[Kel] W. T. B. Kelvin, *Mathematical and Physical Papers*, Cambridge University Press, 1910.

[Ke1] C. Kenig, *Recent progress on boundary-value problems on Lipschitz domains*, Proceedings of Symposia in Pure Mathematics, vol. 43, 1985, pp. 175–205.

[Ke2] C. E. Kenig, *Harmonic analysis techniques for second order elliptic boundary value problems*, vol. 8, CBMS Regional Conference Series in Mathematics, AMS, Providence, RI, 1994.

[Ko] K. Kodaira, *Harmonic fields in Riemannian manifolds (Generalized potential theory)*, Annals of Math. **50** (1949), 587–665.

[KS] J. J. Kohn and D. C. Spencer, *Complex Neumann problems*, Annals of Math. **66** (1957), 89–140.

[KN] H. Kumano-go and M. Nagase, *Pseudo-differential operators with nonregular symbols and applications*, Funkcial Ekvac. **21** (1978), 151–192.

[MP] K. McLeod and R. Picard, *A compact imbedding result on Lipschitz manifolds*, Math. Ann. **290** (1991), 491–508.

[Me] Y. Meyer, *Ondelettes et Opérateures*, vol. II, Hermann, Paris, 1990.

[MD1] D. Mitrea, *Layer potential operators and boundary value problems for differential forms on Lipschitz domains,*, Ph. D. thesis, University of Minnesota, 1996.

[MD2] D. Mitrea, *The method of layer potentials for non-smooth domains with arbitrary topology*, J. Int. Eq. Op. Th. **29** (1997), 320–338.

[MM1] D. Mitrea and M. Mitrea, *Boundary integral methods for harmonic differential forms in Lipschitz domains*, Electronic Research Announcements of the Amer. Math. Soc. **2** (1996), 92–97.

[MM2] D. Mitrea and M. Mitrea, *Uniqueness for inverse conductivity and transmission problems in the class of Lipschitz domain*, Comm. PDE **23** (1998), 1419–1448.

[MMP] D. Mitrea, M. Mitrea and J. Pipher, *Vector potential theory on non-smooth domains in* $\mathbb{R}^3$ *and applications to electromagnetic scattering*, J. Four. Anal. Appl. **3** (1997), 131–192.

[Mi1] M. Mitrea, *Clifford Algebras in Harmonic Analysis and Elliptic Boundary Value Problems on Nonsmooth Domains* (1994), Ph. D. Dissertation, University of South Carolina, Columbia.

[Mi2] M. Mitrea, *Electromagnetic scattering on nonsmooth domains*, Math. Res. Letters **1** (1994), 639–646.

[Mi3] M. Mitrea, *The method of layer potentials in electromagnetic scattering theory on nonsmooth domains*, Duke Math. J. **77** (1995), 111–133.

[MT1] M. Mitrea and M. Taylor, *Boundary layer methods for Lipschitz domains in Riemannian manifolds*, J. Funct. Anal. **163** (1999), 181–251.

[MT2] M. Mitrea and M. Taylor, *Potential theory on Lipschitz domains in Riemannian manifolds: Sobolev-Besov space results and the Poisson problem*, to appear in J. Funct. Anal. (2000).

[Mo1] C. B. Morrey, Jr., *A variational method in the theory of harmonic integrals, II*, Amer. J. of Math. **78** (1956), 137–170.

[Mo2] C. B. Morrey, Jr., *Second order elliptic systems of differential equations*, Contributions to the Theory of Partial Differential Equations (L. Bers et al., ed.), Princeton Univ. Press, Princeton, 1954, pp. 101–159.

[Mo3] C. B. Morrey, Jr., *Multiple Integrals in the Calculus of Variations*, Springer-Verlag, 1966.

[ME1] C. B. Morrey, Jr. and J. Eells, Jr., *A variational method in the theory of harmonic integrals*, Proc. Nat. Acad. Sci. **41** (1955), 391–395.

[ME2] C. B. Morrey, Jr. and J. Eells, Jr., *A variational method in the theory of harmonic integrals I*, Annals of Math. **63** (1956), 91–128.

[Ne] J. Nečas, *Les Méthodes Directes en Théorie des Équations Élliptiques*, Academia, Prague, 1967.

[Pa] R. Palais, *Seminar on the Atiyah-Singer Index Theorem*, vol. 57, Annals of Math. Studies, Princeton University Press, Princeton, New Jersey, 1965.

[PW] L. Payne and H. Weinberger, *New bounds for solutions of second order elliptic partial differential equations*, Pacific J. Math. **8** (1958), 551–573.

[Pi] R. Picard, *An elementary proof for a compact imbedding result in generalized electromagnetic theory*, Math. Z. **187** (1984), 151–164.

[PV] J. Pipher and G. Verchota, *Dilation invariant estimates and the boundary Gårding inequality for higher order elliptic operators*, Annals of Math. **142** (1995), 1–38.

[Sc] G. Schwarz, *Hodge Decomposition-A method for solving boundary value problems*, Lecture Notes in Mathematics, No. 1607, Springer-Verlag, 1995.

[Sh] V. A. Sharafutdinov, *Integral Geometry of Tensor Fields*, VSP, Utrecht, The Netherlands, 1994.

[Sp] D. C. Spencer, *Real and complex operators on manifolds*, Contributions to the theory of Riemann surfaces, Princeton University Press, 1953, pp. 204–227.

[Ta1] M. Taylor, *Pseudodifferential Operators and Nonlinear PDE*, Birkhäuser, Boston, 1991.

[Ta2] M. Taylor, *Partial Differential Equations*, vol. I-III, Springer-Verlag, 1996.

[Te1] N. Teleman, *The Index Theorem for topological manifolds*, Acta Math. **153** (1984), 117-152.

[Te2] N. Teleman, *The index of signature operators on Lipschitz manifolds*, Inst. Hautes Études Sci. Publ. Math. **58** (1983), 39–78.

[Tu] A. W. Tucker, *A boundary-value theorem for harmonic tensors*, Bull. Amer. Math. Soc. **47** (1941), 714.

[Va] F. H. Vasilescu, *Initiere in teoria operatorilor liniari (in Romanian)*, Ed. Tehnica Bucuresti, 1987.

[Ve1] G. Verchota, *Layer potentials and boundary value problems for Laplace's equation in Lipschitz domains*, J. Funct. Anal. **59** (1984), 572–611.

[Ve2] G. Verchota, *Remarks on 2nd order elliptic systems in Lipschitz domains*, Miniconference on operator theory and partial differential equations , North Ryde, 1986, vol. 14, Proc. Center Math. Anal. Austral. Nat. Univ., Austral. Nat. Univ., Canberra, 1986, pp. 303–325.

[VV] G. C. Verchota and A. L. Vogel, *Nonsymmetric systems on nonsmooth planar domains*, Trans. Amer. Math. Soc. **349** (1997), 4501–4535.

[Wa] F. W. Warner, *Foundations of Differentiable Manifolds and Lie Groups*, Springer-Verlag, 1994.

Editorial Information

To be published in the *Memoirs*, a paper must be correct, new, nontrivial, and significant. Further, it must be well written and of interest to a substantial number of mathematicians. Piecemeal results, such as an inconclusive step toward an unproved major theorem or a minor variation on a known result, are in general not acceptable for publication. Papers appearing in *Memoirs* are generally longer than those appearing in *Transactions*, which shares the same editorial committee.

As of November 30, 2000, the backlog for this journal was approximately 10 volumes. This estimate is the result of dividing the number of manuscripts for this journal in the Providence office that have not yet gone to the printer on the above date by the average number of monographs per volume over the previous twelve months, reduced by the number of volumes published in four months (the time necessary for preparing a volume for the printer). (There are 6 volumes per year, each containing at least 4 numbers.)

A Consent to Publish and Copyright Agreement is required before a paper will be published in the *Memoirs*. After a paper is accepted for publication, the Providence office will send a Consent to Publish and Copyright Agreement to all authors of the paper. By submitting a paper to the *Memoirs*, authors certify that the results have not been submitted to nor are they under consideration for publication by another journal, conference proceedings, or similar publication.

Information for Authors

Memoirs are printed from camera copy fully prepared by the author. This means that the finished book will look exactly like the copy submitted.

The paper must contain a *descriptive title* and an *abstract* that summarizes the article in language suitable for workers in the general field (algebra, analysis, etc.). The *descriptive title* should be short, but informative; useless or vague phrases such as "some remarks about" or "concerning" should be avoided. The *abstract* should be at least one complete sentence, and at most 300 words. Included with the footnotes to the paper should be the 2000 *Mathematics Subject Classification* representing the primary and secondary subjects of the article. The classifications are accessible from `www.ams.org/msc/`. The list of classifications is also available in print starting with the 1999 annual index of *Mathematical Reviews*. The Mathematics Subject Classification footnote may be followed by a list of *key words and phrases* describing the subject matter of the article and taken from it. Journal abbreviations used in bibliographies are listed in the latest *Mathematical Reviews* annual index. The series abbreviations are also accessible from `www.ams.org/publications/`. To help in preparing and verifying references, the AMS offers MR Lookup, a Reference Tool for Linking, at `www.ams.org/mrlookup/`. When the manuscript is submitted, authors should supply the editor with electronic addresses if available. These will be printed after the postal address at the end of the article.

Electronically prepared manuscripts. The AMS encourages electronically prepared manuscripts, with a strong preference for AMS-LaTeX. To this end, the Society has prepared AMS-LaTeX author packages for each AMS publication. Author packages include instructions for preparing electronic manuscripts, the *AMS Author Handbook*, samples, and a style file that generates the particular design specifications of that publication series. Though AMS-LaTeX is the highly preferred format of TeX, author packages are also available in AMS-TeX.

Authors may retrieve an author package from e-MATH starting from `www.ams.org/tex/` or via FTP to `ftp.ams.org` (login as `anonymous`, enter username as password, and type `cd pub/author-info`). The *AMS Author Handbook* and the *Instruction Manual* are available in PDF format following the author packages link from `www.ams.org/tex/`. The author package can be obtained free of charge by sending email to `pub@ams.org` (Internet) or from the Publication Division, American Mathematical Society, P.O. Box 6248, Providence, RI 02940-6248. When requesting an author package, please specify AMS-LaTeX or AMS-TeX, Macintosh or IBM (3.5) format, and the publication in which your paper will appear. Please be sure to include your complete mailing address.

Sending electronic files. After acceptance, the source file(s) should be sent to the Providence office (this includes any TeX source file, any graphics files, and the DVI or PostScript file).

Before sending the source file, be sure you have proofread your paper carefully. The files you send must be the EXACT files used to generate the proof copy that was accepted for publication. For all publications, authors are required to send a printed copy of their paper, which exactly matches the copy approved for publication, along with any graphics that will appear in the paper.

TeX files may be submitted by email, FTP, or on diskette. The DVI file(s) and PostScript files should be submitted only by FTP or on diskette unless they are encoded properly to submit through email. (DVI files are binary and PostScript files tend to be very large.)

Electronically prepared manuscripts can be sent via email to `pub-submit@ams.org` (Internet). The subject line of the message should include the publication code to identify it as a Memoir. TeX source files, DVI files, and PostScript files can be transferred over the Internet by FTP to the Internet node `e-math.ams.org` (130.44.1.100).

Electronic graphics. Comprehensive instructions on preparing graphics are available at `www.ams.org/jourhtml/graphics.html`. A few of the major requirements are given here.

Submit files for graphics as EPS (Encapsulated PostScript) files. This includes graphics originated via a graphics application as well as scanned photographs or other computer-generated images. If this is not possible, TIFF files are acceptable as long as they can be opened in Adobe Photoshop or Illustrator. No matter what method was used to produce the graphic, it is necessary to provide a paper copy to the AMS.

Authors using graphics packages for the creation of electronic art should also avoid the use of any lines thinner than 0.5 points in width. Many graphics packages allow the user to specify a "hairline" for a very thin line. Hairlines often look acceptable when proofed on a typical laser printer. However, when produced on a high-resolution laser imagesetter, hairlines become nearly invisible and will be lost entirely in the final printing process.

Screens should be set to values between 15% and 85%. Screens which fall outside of this range are too light or too dark to print correctly. Variations of screens within a graphic should be no less than 10%.

Inquiries. Any inquiries concerning a paper that has been accepted for publication should be sent directly to the Electronic Prepress Department, American Mathematical Society, P. O. Box 6248, Providence, RI 02940-6248.

Editors

This journal is designed particularly for long research papers, normally at least 80 pages in length, and groups of cognate papers in pure and applied mathematics. Papers intended for publication in the *Memoirs* should be addressed to one of the following editors. In principle the Memoirs welcomes electronic submissions, and some of the editors, those whose names appear below with an asterisk (*), have indicated that they prefer them. However, editors reserve the right to request hard copies after papers have been submitted electronically. Authors are advised to make preliminary email inquiries to editors about whether they are likely to be able to handle submissions in a particular electronic form.

Algebra to CHARLES CURTIS, Department of Mathematics, University of Oregon, Eugene, OR 97403-1222 email: cwc@darkwing.uoregon.edu

Algebraic geometry and commutative algebra to LAWRENCE EIN, Department of Mathematics, University of Illinois, 851 S. Morgan (M/C 249), Chicago, IL 60607-7045; email: ein@uic.edu

Algebraic topology and cohomology of groups to STEWART PRIDDY, Department of Mathematics, Northwestern University, 2033 Sheridan Road, Evanston, IL 60208-2730; email: priddy@math.nwu.edu

Combinatorics and Lie theory to SERGEY FOMIN, Department of Mathematics, University of Michigan, Ann Arbor, Michigan 48109-1109; email: fomin@math.lsa.umich.edu

Complex analysis and complex geometry to DUONG H. PHONG, Department of Mathematics, Columbia University, 2990 Broadway, New York, NY 10027-0029; email: dp@math.columbia.edu

*__Differential geometry and global analysis__ to LISA C. JEFFREY, Department of Mathematics, University of Toronto, 100 St. George St., Toronto, ON Canada M5S 3G3; email: jeffrey@math.toronto.edu

*__Dynamical systems and ergodic theory__ to ROBERT F. WILLIAMS, Department of Mathematics, University of Texas, Austin, Texas 78712-1082; email: bob@math.utexas.edu

Geometric topology, knot theory, hyperbolic geometry, and general topoogy to JOHN LUECKE, Department of Mathematics, University of Texas, Austin, TX 78712-1082; email: luecke@math.utexas.edu

Harmonic analysis, representation theory, and Lie theory to ROBERT J. STANTON, Department of Mathematics, The Ohio State University, 231 West 18th Avenue, Columbus, OH 43210-1174; email: stanton@math.ohio-state.edu

*__Logic__ to THEODORE SLAMAN, Department of Mathematics, University of California, Berkeley, CA 94720-3840; email: slaman@math.berkeley.edu

Number theory to MICHAEL J. LARSEN, Department of Mathematics, Indiana University, Bloomington, IN 47405; email: larsen@math.indiana.edu

Operator algebras and functional analysis to BRUCE E. BLACKADAR, Department of Mathematics, University of Nevada, Reno, NV 89557; email: bruceb@math.unr.edu

*__Ordinary differential equations, partial differential equations, and applied mathematics__ to PETER W. BATES, Department of Mathematics, Brigham Young University, 292 TMCB, Provo, UT 84602-1001; email: peter@math.byu.edu

*__Partial differential equations and applied mathematics__ to BARBARA LEE KEYFITZ, Department of Mathematics, University of Houston, 4800 Calhoun Road, Houston, TX 77204-3476; email: keyfitz@uh.edu

*__Probability and statistics__ to KRZYSZTOF BURDZY, Department of Mathematics, University of Washington, Box 354350, Seattle, Washington 98195-4350; email: burdzy@math.washington.edu

*__Real and harmonic analysis and geometric partial differential equations__ to WILLIAM BECKNER, Department of Mathematics, University of Texas, Austin, TX 78712-1082; email: beckner@math.utexas.edu

All other communications to the editors should be addressed to the Managing Editor, WILLIAM BECKNER, Department of Mathematics, University of Texas, Austin, TX 78712-1082; email: beckner@math.utexas.edu.

Selected Titles in This Series

(Continued from the front of this publication)

683 **Serge Bouc,** Non-additive exact functors and tensor induction for Mackey functors, 2000

682 **Martin Majewski,** ational homotopical models and uniqueness, 2000

681 **David P. Blecher, Paul S. Muhly, and Vern I. Paulsen,** Categories of operator modules (Morita equivalence and projective modules, 2000

680 **Joachim Zacharias,** Continuous tensor products and Arveson's spectral C^*-algebras, 2000

679 **Y. A. Abramovich and A. K. Kitover,** Inverses of disjointness preserving operators, 2000

678 **Wilhelm Stannat,** The theory of generalized Dirichlet forms and its applications in analysis and stochastics, 1999

677 **Volodymyr V. Lyubashenko,** Squared Hopf algebras, 1999

676 **S. Strelitz,** Asymptotics for solutions of linear differential equations having turning points with applications, 1999

675 **Michael B. Marcus and Jay Rosen,** Renormalized self-intersection local times and Wick power chaos processes, 1999

674 **R. Lawther and D. M. Testerman,** A_1 subgroups of exceptional algebraic groups, 1999

673 **John Lott,** Diffeomorphisms and noncommutative analytic torsion, 1999

672 **Yael Karshon,** Periodic Hamiltonian flows on four dimensional manifolds, 1999

671 **Andrzej Rosłanowski and Saharon Shelah,** Norms on possibilities I: Forcing with trees and creatures, 1999

670 **Steve Jackson,** A computation of δ^1_5, 1999

669 **Seán Keel and James M^c^Kernan,** Rational curves on quasi-projective surfaces, 1999

668 **E. N. Dancer and P. Poláčik,** Realization of vector fields and dynamics of spatially homogeneous parabolic equations, 1999

667 **Ethan Akin,** Simplicial dynamical systems, 1999

666 **Mark Hovey and Neil P. Strickland,** Morava K-theories and localisation, 1999

665 **George Lawrence Ashline,** The defect relation of meromorphic maps on parabolic manifolds, 1999

664 **Xia Chen,** Limit theorems for functionals of ergodic Markov chains with general state space, 1999

663 **Ola Bratteli and Palle E. T. Jorgensen,** Iterated function systems and permutation representation of the Cuntz algebra, 1999

662 **B. H. Bowditch,** Treelike structures arising from continua and convergence groups, 1999

661 **J. P. C. Greenlees,** Rational S^1-equivariant stable homotopy theory, 1999

660 **Dale E. Alspach,** Tensor products and independent sums of $\mathcal{L}_p$-spaces, $1 < p < \infty$, 1999

659 **R. D. Nussbaum and S. M. Verduyn Lunel,** Generalizations of the Perron-Frobenius theorem for nonlinear maps, 1999

658 **Hasna Riahi,** Study of the critical points at infinity arising from the failure of the Palais-Smale condition for n-body type problems, 1999

657 **Richard F. Bass and Krzysztof Burdzy,** Cutting Brownian paths, 1999

656 **W. G. Bade, H. G. Dales, and Z. A. Lykova,** Algebraic and strong splittings of extensions of Banach algebras, 1999

655 **Yuval Z. Flicker,** Matching of orbital integrals on $GL(4)$ and $GSp(2)$, 1999

654 **Wancheng Sheng and Tong Zhang,** The Riemann problem for the transportation equations in gas dynamics, 1999

For a complete list of titles in this series, visit the AMS Bookstore at **www.ams.org/bookstore/**.